# SpringerBriefs in Computer Science

More information about this series at http://www.springer.com/series/10028

Yin Zhang • Min Chen

# Cloud Based 5G Wireless Networks

 Springer

Yin Zhang
School of Information and Safety
Engineering
Zhongnan University of Economics
and Law
Wuhan, Hubei, China

Min Chen
School of Computer Science
and Technology
Huazhong University of Science
and Technology
Wuhan City, China

ISSN 2191-5768          ISSN 2191-5776  (electronic)
SpringerBriefs in Computer Science
ISBN 978-3-319-47342-0          ISBN 978-3-319-47343-7  (eBook)
DOI 10.1007/978-3-319-47343-7

Library of Congress Control Number: 2016954921

Printed on acid-free paper

This Springer imprint is published by Springer Nature
The registered company is Springer International Publishing AG
The registered company address is: Gewerbestrasse 11, 6330 Cham, Switzerland

# Preface

In recent years, information communication and computation technologies are deeply converging, and various wireless access technologies have been successful in deployment. It can be predicted that the upcoming fifth-generation mobile communication technology (5G) can no longer be defined by a single business model or a typical technical characteristic. 5G is a multi-service and multi-technology integrated network, meeting the future needs of a wide range of big data and the rapid development of numerous businesses, and enhancing the user experience by providing smart and customized services. In this book, we introduce the general background of 5G wireless networks and review related technologies, such as cloud-based networking, cloud platform for networking, definable networking, green wireless networks, which are capable of providing a virtualized, reconfigurable, smart wireless network.

We are grateful to Dr. Xuemin (Sherman) Shen, the SpringerBriefs Series Editor on Wireless Communications. This book would not be possible without his kind support during the process. Thanks also to the Springer Editors and Staff, all of whom did their usual excellent job in getting this monograph published.

This work was supported by China National Natural Science Foundation (No. 61572220).

Wuhan, China                                                      Yin Zhang
September 2016                                                    Min Chen

# Contents

**1 Introduction** ......................................................... 1
  1.1   The Development of Wireless Networks ............................. 1
  1.2   5G Wireless Networks ..................................................... 3
  References ...................................................................... 7

**2 Cloud-Based Networking** .......................................... 9
  2.1   Network Foundation Virtualization ..................................... 9
      2.1.1   Development Status of NFV ..................................... 11
      2.1.2   Technical Issues of NFV ........................................ 13
  2.2   Cloud Radio Access Networks ......................................... 15
  2.3   Mobile Cloud Networking ............................................... 17
  References ...................................................................... 18

**3 Cloud Platform for Networking** ................................. 21
  3.1   OpenNebule ............................................................... 21
  3.2   OpenStack .................................................................. 23
  3.3   OpenDayLight .............................................................. 27
  3.4   Virtual Machine Migration .............................................. 28
      3.4.1   P2V .................................................................. 28
      3.4.2   V2V .................................................................. 29
      3.4.3   V2P .................................................................. 30
  References ...................................................................... 30

**4 Definable Networking** .............................................. 33
  4.1   Caching .................................................................... 33
  4.2   Mobile Content Distribution Network .................................. 34
  4.3   Software-Defined Mobile Network ..................................... 36
      4.3.1   SDN Architecture ................................................. 38
      4.3.2   The Critical Techniques for Data Layer ...................... 40
      4.3.3   The Critical Techniques for Control Layer ................... 43
      4.3.4   SDN-Based Application .......................................... 51

4.4   Networking as a Service ................................................  54
    4.4.1   Create a Virtual Network Segment...........................  54
    4.4.2   Integration of NaaS and WAN ...............................  55
    4.4.3   Advantage of NaaS ..........................................  55
    References .................................................................  55

5  **Green Wireless Networks** ....................................................  59
    5.1   Background..............................................................  59
    5.2   Cognitive SDN for Green Wireless Networks .........................  61
        5.2.1   Cognitive SDN Architecture and Technology ...................  61
        5.2.2   Green Wireless Network Architecture Based
                on Cognitive SDN ..............................................  63
    5.3   SDN-Based Energy Efficiency Optimization for RAN................  64
        5.3.1   Separation Between Control and Data ........................  64
        5.3.2   Separation Between Uplink and Downlink ....................  65
        5.3.3   Elastic Wireless Resources Matching..........................  65
    5.4   SDN-Based Green Wireless Networks Fusion .........................  66
    References .................................................................  66

6  **5G-Related Projects** .........................................................  69
    6.1   METIS.................................................................  69
    6.2   Multi-hop Cellular Networks ..........................................  72
    6.3   T-NOVA ...............................................................  73
    6.4   iJOIN .................................................................  75
    6.5   NUAGE................................................................  75
    References .................................................................  77

7  **5G-Based Applications** ......................................................  79
    7.1   RAN Sharing ..........................................................  79
    7.2   Multi-Operator Core Network..........................................  81
    7.3   Fixed Mobile Convergence .............................................  82
    7.4   Small Cells ............................................................  83
    7.5   Other Applications .....................................................  83
    References .................................................................  84

8  **Conclusion** .................................................................  85
    References .................................................................  86

# Acronyms

| | |
|---|---|
| 3GPP | 3rd Generation Partnership Project |
| 4G | The fourth generation mobile cellular communication system |
| 5G | The fifth-generation mobile communication |
| AMQP | Advanced Message Queue Protocol |
| Amazon EC2 | Amazon Elastic Compute Cloud |
| API | Application Programming Interface |
| ARFCN | Absolute Radio Frequency Channel Number |
| AVI | Architecture of the virtualization infrastructure |
| BBU | Bandwidth Based Unit |
| BGP-LS | Border Gateway Protocol Link-State |
| BSC | Base Station Controller |
| BSS | Business Support Systems |
| CC | Cloud Controller |
| CDMA | Code division multiple access |
| CDN | Content delivery network |
| CDPI | Control-data-plane interface |
| Cloud-RAN | Cloud Radio Access Networks |
| COTS | Commercial off-the-shelf |
| D2D | Device-to-device |
| DC | Data center |
| DG CONNECT | Directorate General for Communications Networks, Content & Technology |
| DHCP | Dynamic Host Configuration Protocol |
| EMS | Element management system |
| eNobeB | Evolved Node B |
| EPC | Evolved Packet Core |
| ETSI | European Telecommunication Standards Institute |
| EVE | Evolution and Ecosystem |
| EXR | Exclusive routing |
| FDMA | Frequency division multiple access |
| FIS | Flow instruction set |

| | |
|---|---|
| FMC | Fixed mobile convergence |
| ForCES | Forwarding and Control Element Separation |
| FRP | Functional reactive programming |
| GGSN | Gateway GPRS Support Node |
| GPU | Graphics Processing Unit |
| HFT | Hierarchical Flow Tables |
| HSPA+ | Evolved High Speed Packet Access |
| IAAS | Infrastructure as a service |
| ICT | Information and Communication Technology |
| IETF | Internet Engineering Task Force |
| IFA | Interfaces and Architecture |
| IGP | Interior Gateway Protocol |
| IMT-A | International Mobile Telecommunication-Advanced |
| IoT | Internet of Things |
| IoV | Internet of Vehicles |
| IP | Internet Protocol |
| IRTF | Internet Research Task Force |
| ISG | Industry Specification Group |
| IT | Information technology |
| ITU | International Telecommunications Union |
| ITU-T | ITU Telecommunication Standardization Sector |
| KVM | Kernel-Based Virtual Machine |
| LGW | Local Gateway |
| LTE | Long term evolution |
| MAC | Media Access Control |
| MANO | Management & orchestration |
| MBMS | Multimedia Broadcast Multicast Services |
| MCDN | Mobile Content Distribution Network |
| MCN | Multi-hop Cellular Networks |
| MCN | Mobile cloud networking |
| METIS | Mobile and wireless communications Enablers for the Twenty-twenty Information Society |
| MIMO | Multiple-input multiple-output |
| MMC | Massive machine communication |
| MME | Mobility Management Entity |
| MN | Moving network |
| MOCN | Multi-operator Core Network |
| NaaS | Networking as a Service |
| NBI | Northbound interface |
| NE | Network element |
| NFaaS | Network Functions as-a-Service |
| NFV | Network Foundation Virtualization |
| NFVI | NFV infrastructure |
| NFVO | NFV Orchestrator |
| NIB | Network information base |

| | |
|---|---|
| NMS | Network management system |
| NP | Network processor |
| NUAGE | Nuage Virtualized Services Platform |
| OFDM | Orthogonal frequency division multiplexing |
| ONF | Open Networking Foundation |
| ONOS | Open Network Operating System |
| OSGi | Open Services Gateway initiative |
| OSS | Operation support system |
| P&P | Performance & Portability |
| P2V | Physical-to-Virtual |
| PBX | Private branch exchange |
| PGW | Packet Data Network Gateway |
| POC | Proof of concept |
| QoE | Quality of Experience |
| QoS | Quality of Service |
| R&A | Reliability & Availability |
| RAN | Radio access networks |
| RANaaS | RAN-as-a-Service |
| REL | Reliability |
| REST | Representational State Transfer |
| RNC | Radio Network Controller |
| RRU | Remote Radio Unit |
| RSSI | Received signal strength indicator |
| SA | Software architecture |
| SAL | Service abstraction layer |
| SCN | Single-hop Cellular Networks |
| SDN | Software defined network |
| SDNGR | Software-Defined Networking Research Group |
| SEC | Security |
| SGSN | Serving GPRS Support Node |
| SGW | Serving Gateway |
| SM | Service Manager |
| SNA | Shared Network Arca |
| SO | Service Orchestrator |
| SON | Self-Organization Network |
| T-NOVA | Network function as-a-service over virtualized infrastructures |
| TDMA | Time division multiple access |
| TMSI | Temporary Mobile Subscriber Identifier |
| TSC | Technical Steering Committee |
| TST | Testing, Experimentation and Open Source |
| UDN | Ultra dense networking |
| URC | Ultra reliable communication |
| V2V | Virtual-to-Virtual |
| V2P | Virtual-to-Physical |
| vCDN | virtual Content Distribution Network |

| vCPE | virtual Customer Premise Equipment |
| vEPC | virtualized Evolved Packet Core |
| vIMS | virtual IP Multimedia Subsystem |
| VLAN | Virtual local area network |
| VM | Virtual machine |
| VMM | Virtual Machine Monitor |
| VNF | Virtual Network Function |
| VNF-FG | VNF Forwarding Graph |
| VNFC | VNF Component |
| VNFL | VNF Link |
| VNS | Virtualized network services |
| VPN | Virtual private network |
| VXLAN | Virtual Extensible LAN |
| WG | Working group |
| WLAN | Wireless local area network |
| WRC | World Radio Communication Conference |

# Chapter 1
# Introduction

**Abstract** In recent years, information communication and computation technologies are deeply converging, and various wireless access technologies have been successful in deployment. It can be predicted that the upcoming fifth-generation mobile communication technology (5G) can no longer be defined by a single business model or a representative technical characteristic. 5G is a multi-service and multi-technology integrated network, meeting the future needs of a wide range of big data and the rapid development of numerous businesses, and enhancing the user experience by providing intelligent and personalized services.

## 1.1 The Development of Wireless Networks

Wireless networks have been rapidly developing in the recent 20 years. They have brought a huge impact to all aspects of people's lifestyles in terms of work, social, and economy. Human society has entered the information era with the support of big data. The demand for advanced technologies to support future applications and services in all aspects of people's living is continuously increasing. Moreover, with the rapid development of wearable devices, Internet of Things (IoT), Internet of Vehicles (IoV), etc., both numbers and types of smart devices accessing to wireless networks will overwhelm the ability of existing networks.

According to "Global Mobile Data Traffic Forecast Update 2014–2019 White Paper" by Cisco [3], the global mobile data traffic at the end of 2013 reached 1.5 exabytes, increased by 81 % from 2012, while the mobile data traffic of 2014 was nearly as 18 times as the total annual amount of Internet data in 2000. Moreover, the global mobile data traffic from 2014 to 2020 is continuing to grow exponentially, and it estimates that the demand for data capacity will grow 1000-fold in the next 10 years. Especially, the mobile data generated by cellular networks will account for more than 60 %, and the mobile wireless networks traffic in 2020 will be as 500 times as 2010 [8]. The explosion of mobile data is bringing the following challenges for the wireless networks:

© The Author(s) 2016
Y. Zhang, M. Chen, *Cloud Based 5G Wireless Networks*, SpringerBriefs
in Computer Science, DOI 10.1007/978-3-319-47343-7_1

- **Connectivity capacity**: Traditional communication technologies mainly provide human-to-human communication. With the rise of IoT and other related technologies, more devices can access to the networks that the increasing needs for human-to-device and device-to-device communication should be satisfied. Thus, the fifth-generation mobile communication technology (5G) is expected to provide a ubiquitous solution to connect everything any time, any where.
- **Network performance**: Due to more novel applications accessing to the mobile networks, people expect to easily and rapidly access to a rich variety of information. In any environment, users can conveniently be provided real-time access to multimedia resources, and other useful information through 5G networks.
- **Resource optimization**: The Quality of Service (QoS) of traditional communication technologies is often improved by upgrading the hardware and other infrastructures. However, this approach needs more cost and easily cases a waste of resources. 5G networks are expected to intelligently identify the communication scenarios, dynamically allocate network resources and provide considerable connectivity and network performance on demand, and improve the efficiency of existing resources.

Surge in broadband mobile data services demands for high-throughput, low-latency data transmission, so promoting communication operators laid more and more intensive base station equipment to meet the coverage capacity of user groups in different regions and hotspots, while the traditional cellular network architecture in the long-term evolution is also becoming heterogeneous, complex, and intensive. In November 2010, the International Telecommunication Union (ITU) approved the International Mobile Telecommunication-Advanced (IMT-A) international standard. Since then, the fourth generation mobile cellular communication system (4G) based on this standard is widely implemented all over the world, and the 4G Long-Term Evolution (LTE) network is one of the most representative technique [1]. However, people are continuing to expect more advanced communications, so the innovation of mobile network technology will never stop. So far, though 4G has been matured, with the publishing of IMT-2020 international standard, 5G is improving rapidly. Compared to 4G-LTE, 5G is expected to support 10 times the present data capacity, 10–100 times the present number and speed of available connection, 10 times the present battery life time and one-fifth the present delay. In 2015, the 3rd Generation Partnership Project (3GPP) published the main technical requirements for performance comparison between IMT-A and IMT-2020, from which it is obvious that the future 5G will be a comprehensive, profound, and advanced technological.[1,2]

---

[1] Third Generation Partnership Project (3GPP), http://www.3gpp.org/.

[2] Third Generation Partnership Project 2 (3GPP2), http://www.3gpp2.org/.

## 1.2  5G Wireless Networks

5G is gradually becoming the new hotspot of academia and industry [6]. It is expected that 5G will be the leading mobile communication technology after 2020 to meet the information requirement of the human society by interconnecting the wireless world without barriers [13]. With the enhancement of bandwidth and capacity of wireless mobile communication systems and the rapid development of the applications of mobile networks, the IoT and mobile wireless networks for personal usage and business will be evolved with fundamental ecological changes. Wireless communication, computer, and information technology will be closely and deeply interworked, and the novel hardware and software will be rapidly improved to support the development of 5G industry.

Although 5G has been proposed with a basic idea and prototype, it meets a series of key technical challenges and tremendous obstacles. In particular, there is a profound contradiction between the growing demand for wireless communication services and the increasingly complex heterogeneous network environment, which causes greater resource and energy consumption for improving network capacity. Since the twentieth century, with the increasingly prominent global warming, climate anomalies, energy crisis, and other related issues, the development of low-carbon economy has become the consensus of the human community, and it is a fundamental requirement to develop a sustainable, resource-optimized and energy efficient green communications technology, and network infrastructure. As shown in Fig. 1.1, it effectively reduces the resources and energy consumption, and improves resource and energy efficiency in accordance with the principles and re-examine the design of future wireless communication networks. Therefore, 5G is a significant research and innovation, and the development of 5G technology is the only way for wireless networks evolution.

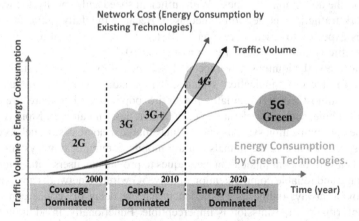

**Fig. 1.1** Sustainable Green Communications Evolution for 5G by 2020 (*Source*: UEB—Labex COMIN Labs, 2014)

In particular, Shannon channel capacity theory suggests that there is a linear relationship between the capacity and bandwidth, while the relationship between the capacity and the power is logarithmic. The fundamental theorem reveals the existence of a compromise between the required power and spectral bandwidth, which means that it is available to decrease the energy consumption by increasing the effective bandwidth within a limited capacity, and vice versa. It has been proved that if mobile operators can use the novel wireless access technology to dynamically manage their licensed spectrum and make full use of the available spectrum resources to improve the utilization and the energy efficiency, the system can save about 50 % energy consumption. In addition, several statistics also show that the substantial increase in energy consumption, as well as spectrum resource, is another bottleneck for the future as the main heterogeneous wireless communication network development. Specifically, if the existing solutions are desperate used to improve the system capacity, coverage, spectral efficiency, and other properties, it will significantly increase the network energy consumption, even delay or hinder the sustainable development of future heterogeneous wireless communication network.

Furthermore, with the improvement of the bandwidth and capacity for wireless networks, the mobile applications for individuals and industries are rapidly developed, and the mobile communication related industrial ecology will gradually evolved. 5G is no longer just a air interface technology with higher rate, greater bandwidth, greater capacity, but also a intelligent network for business applications and user experience. Specifically, 5G should achieve the following objectives:

- **Sufficiency**: The user's reliance on mobile applications require the next-generation wireless mobile networks to provide users with enough speed and capacity. Predictably, major mobile terminals demand for a transfer rate of more than 10 Mbps to support full high-definition video transmission, some special scenarios demand for more than 100 Mbps to support ultra high-definition video transmission, and even some particular devices demand for more than 10 Gbps to support the holographic business. With sufficient speed and capacity, the wireless network traffic is expected to be increased. Because the daily traffic from each user is expected to reach more than 1 GBytes, while the special terminals with high traffic volume demand for even more than 10 GBytes.
- **Friendliness**: Ubiquitous coverage and stable quality are the basic requirements for the communication systems. The existing mobile communication systems almost cover essentially the entire population, but there are many coverage hole, such as wilderness, ocean, Antarctica, and aircraft. Moreover, the mobile communication systems may be unavailable in some cases, such as on the high-speed rail and in the tunnels. Future mobile communication systems must include a various communication techniques to provide the users with ubiquitous coverage and reliable communication quality. 5G wireless networks are expected to provide always-online user experience that the delay of service connection and information transmission is imperceptible. Functionally, in addition to the basic communication capabilities and various multimedia applications, more comprehensive applications are provided convenience and efficiency of working and life.

- **Accessibility**: Although 5G includes various complex techniques, from the user's point of view, it is a simple and convenient approach including the following advantages: (1) access technology is transparent to the users, while the network and devices switching are seamless and smooth; (2) the connection between multiple wireless devices is convenient and compatible; (3) the mobile terminals are portable, especially the wearable devices; and (4) the interface to various applications and services are unified.
- **Economy**: This is mainly reflected in two aspects: (1) Although the network traffic continues to increase, the tariff per bit is greatly reduced and it will be even lower. (2) The investment in infrastructure is reduced, while the network resource utilization is improved through dynamic allocation in order to improving the QoS.
- **Personality**: Future communication is a people-oriented, user-experience-centric system that the service is available to be customized by users according to their individual preferences. Specifically, according to the user's preference, network and physical environment, the providers can provide the optimal network access and personalized recommendation.

As shown in Fig. 1.2, the mobile communication develops from the first generation of mobile communication systems (1G) to the fourth generation mobile communication system (4G), while the development of each generation has the operational capacity and representative technology, such as analog cellular technology of 1G; time division multiple access (TDMA), frequency division multiple access (FDMA), and other digital cellular technology supporting voice communication for 2G; code division multiple access (CDMA) supporting data and multimedia services for 3G; orthogonal frequency division multiplexing (OFDM) and multiple-input multiple-output (MIMO) supporting broadband data and mobile networks for 4G. In recent years, the rapid development of integrated circuit

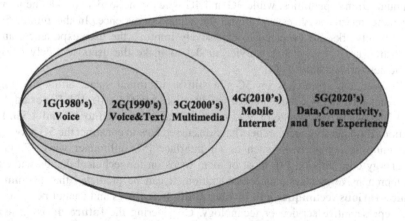

**Fig. 1.2** Service development from 1G to 5G (*Source*: Datang Wireless Mobile Innovation Center, December 2013)

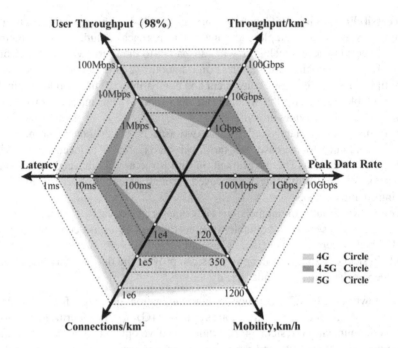

**Fig. 1.3** The performance comparison between 4G, 4.5G, and 5G (*Source*: Datang Wireless Mobile Innovation Center, December 2013)

technology, communication system, and terminal capabilities deeply integrates the communication and computer techniques, various wireless access technologies are developed and wildly implemented. From the perspective of consumers, the initial 1G and 2G wireless networks provide the users with the foundational communication capabilities, while 3G and 4G wireless networks provide the users with more mobile services and higher broadband experience. In the future, 5G wireless networks are expected to extensively improve the user experience, and establish a novel user-centric service model to make the users to freely enjoy mobile networking services [4].

However, the researches on 5G are still at its initial stage, although some documents have defined the technical specifications of 5G [2, 10, 12]. For example, Fig. 1.3 shows the advantage of 5G in performance, compared to 4G and 4.5G. In addition, although some researchers have discussed how to construct the 5G network from multiple perspectives, such as air interface [5], millimeter wave [7, 11], and energy consumption [9], most of them focus on the technical details without a comprehensive and systemic consideration. It can be predicted that **5G must include various techniques and involve multiple features** and cannot be defined by a representative service or technology. Considering the **future development trend** of computer, networking, and communication technologies, 5G will be a virtualized, definable, green mobile communication system providing cloud-based wireless network infrastructure.

# References

1. A. Ghosh, N. Mangalvedhe, R. Ratasuk, B. Mondal, M. Cudak, E. Visotsky, T.A. Thomas, J.G. Andrews, P. Xia, H.S. Jo, et al., Heterogeneous cellular networks: from theory to practice. IEEE Commun. Mag. **50**(6), 54–64 (2012)
2. A. Gohil, H. Modi, S.K. Patel, 5G technology of mobile communication: a survey, in *2013 International Conference on Intelligent Systems and Signal Processing (ISSP)* (IEEE, Vallabh Vidhyanagar, Anand, 2013), pp. 288–292
3. C.V.N. Index, Global mobile data traffic forecast update 2014–2019 white paper, Feb 2015. See: http://www.cisco.com/c/en/us/solutions/collateral/service-provider/visual-networking-index-vni/white_paper_c11-520862.html
4. T. Janevski, 5G mobile phone concept, in *2009 6th IEEE Consumer Communications and Networking Conference* (IEEE, Las Vegas, 2009), pp. 1–2
5. S.G. Larew, T.A. Thomas, M. Cudak, A. Ghosh, Air interface design and ray tracing study for 5G millimeter wave communications, in *2013 IEEE Globecom Workshops (GC Wkshps)* (IEEE, Atlanta, 2013), pp. 117–122
6. Q.C. Li, H. Niu, A.T. Papathanassiou, G. Wu, 5G network capacity: key elements and technologies. IEEE Veh. Technol. Mag. **9**(1), 71–78 (2014)
7. G.R. MacCartney, J. Zhang, S. Nie, T.S. Rappaport, Path loss models for 5G millimeter wave propagation channels in urban microcells, in *2013 IEEE Global Communications Conference (GLOBECOM)* (IEEE, Atlanta, 2013), pp. 3948–3953
8. T. Nakamura, S. Nagata, A. Benjebbour, Y. Kishiyama, T. Hai, S. Xiaodong, Y. Ning, L. Nan, Trends in small cell enhancements in LTE advanced. IEEE Commun. Mag. **51**(2), 98–105 (2013)
9. M. Olsson, C. Cavdar, P. Frenger, S. Tombaz, D. Sabella, R. Jäntti, 5GrEEn: towards green 5G mobile networks, in *The 9th IEEE International Conference on Wireless and Mobile Computing, Networking and Communications (WiMob 2013), International Workshop on the Green Optimized Wireless Networks (GROWN 2013), IEEE Conference Proceedings*, Lyon, 7th October, 2013, pp. 212–216
10. M.S. Pandey, M. Kumar, A. Panwar, I. Singh, A survey: wireless mobile technology generations with 5G. Int. J. Eng. **2**(4), 33–37 (2013)
11. T.S. Rappaport, S. Sun, R. Mayzus, H. Zhao, Y. Azar, K. Wang, G.N. Wong, J.K. Schulz, M. Samimi, F. Gutierrez, Millimeter wave mobile communications for 5G cellular: it will work! IEEE Access **1**, 335–349 (2013)
12. A. Tudzarov, T. Janevski, Functional architecture for 5G mobile networks. Int. J. Adv. Sci. Technol. **32**, 65–78 (2011)
13. L.-C. Wang, S. Rangapillai, A survey on green 5G cellular networks, in *2012 International Conference on Signal Processing and Communications (SPCOM)* (IEEE, Bangalore, 2012), pp. 1–5

# Chapter 2
# Cloud-Based Networking

**Abstract** Cloud networking is an novel approach for building and managing secure private networks over the public Internet through the cloud computing infrastructure. In cloud networking, the traditional network functions and services including connectivity, security, management, and control are pushed to the cloud and published as services, such as Network Foundation Virtualization, Cloud Radio Access Networks, and Mobile Cloud Networking (MCN).

## 2.1 Network Foundation Virtualization

In simplest terms, Network Foundation Virtualization (NFV)[1] is used to migrate the telecommunication equipment from specialized platform to universal x86-based commercial off-the-shelf (COTS) servers. The current telecom networking devices are deployed by the private platforms, within which all the network elements are closed boxes, which cannot utilize the hardware resources mutually. Therefore, the capacity expansion of each devices relies on the additional hardware, while the hardware resources lie idle after the capacity reduction, which is quite time-consuming with poor elasticity and high cost. Through NFV, all the network elements are transformed into independent applications that can be flexibly deployed on a unified platform based on a standard server, storage, and exchange mechanism. As shown in Fig. 2.1, with the decoupled software and hardware, the capacity of every application is available to be expanded rapidly through increasing the virtual resources, and vice versa, which has enhanced the elasticity of the network dramatically.

The technological foundation of NFV is the cloud computing and virtualization techniques in Information Technology (IT) industry. Through the virtualization techniques, the universal resources of computing, storage, and networking provided by COSTS can decompose into a variety of virtual recourses for the use of upper applications. At the same time, the application and hardware are decoupled through the virtualization techniques, while the supply speed of resources has shortened from a few days to a few minutes. Through the cloud computing technology, the

---

[1]http://www.etsi.org/technologies-clusters/technologies/nfv.

© The Author(s) 2016

Y. Zhang, M. Chen, *Cloud Based 5G Wireless Networks*, SpringerBriefs
in Computer Science, DOI 10.1007/978-3-319-47343-7_2

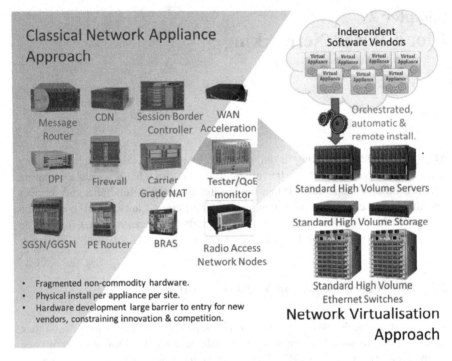

**Fig. 2.1** NFV vision (*Source*: ETSI)

flexible expansion and reduction of the applications is accomplished for contributing to the matching of resources and business load, which does not only improve the resource utilization rate but also ensure the system response speed. Specifically, the deployment of NFV brings the following advantages:

- The purchasing, operation and maintenance costs, and energy consumption of the operators are reduced.
- The business deployment is accelerated, while the innovation cycle is decreased. Specifically, the efficiency of testing and integration are improved, the development cost is reduced, and the conventional hardware deployment is replaced with the quick software installation.
- Network applications support multi-version and multi-tenant to enable the different applications, users, tenants sharing a unified platform, so the network sharing is possible.
- The personalized service of different physical domains and user groups are available, while the service modules can be rapidly expanded.
- The network is open, and the business innovation is able to cause new potential profit increasing point.

### 2.1.1 Development Status of NFV

Since founded in October 2012, the European Telecommunication Standards Institute Industry Specification Group for Network Functions Virtualization (ETSI ISG NFV) develops quickly, which has held six plenary sessions and includes the following works:

- Technical Steering Committee (TSC): takes charge of the overall operating of ETSI ISG NFV;
- Architecture of the Virtualization Infrastructure (AVI): takes charge of the architecture of the virtualization infrastructure;
- Management and Orchestration (MANO): takes charge of management and orchestration;
- Software Architecture (SA): takes charge of software architecture;
- Reliability and Availability (R&A): takes charge of reliability and availability;
- Performance and Portability (P&P): takes charge of performance and portability;
- Security: takes charge of security.

Especially, four overall standards, i.e., Use Cases, Architecture Framework, Terminology for Main Concepts in NFV, and Virtualization Requirements, are finalized by TSC, including five working group (WG) under TSC: Evolution and Ecosystem (EVE), Interfaces and Architecture (IFA), Testing, Experimentation, and Open Source (TST), Security (SEC), and Reliability (REL).

Compared with the current network architecture including independent business network and operation support system (OSS), NFV is deconstructed vertically and horizontally. According to NFV architecture illustrated in Fig. 2.2, from the vertical the network consists of the following three layers: NFV infrastructure (NFVI), Virtual Network Functions (VNFs), and Operation&Business Support Systems (OSS&BSS).

- **NFVI** is a resource pool, from the perspective of cloud computing. The mappings of NFVI on physical infrastructures are some geographically distributed data centers connected by the high-speed communication network.
- **VNFs** correspond with various telecommunication service networks. Each physical network element maps with a VNF. The needed resources fall into virtual computing/storage/exchange resources hosted by NFVI. Interfaces adopted by NFVI are still signaling interfaces defined by the traditional network. Moreover, it still adopts Network Element, Element Management System, and Network Management System (NE-EMS-NMS) framework as its service network management system.
- **OSS&BSS** is the operation support layer needing to make necessary revising and adjusting for its virtualization.

By the horizontal view, NFV includes services network and management and orchestration:

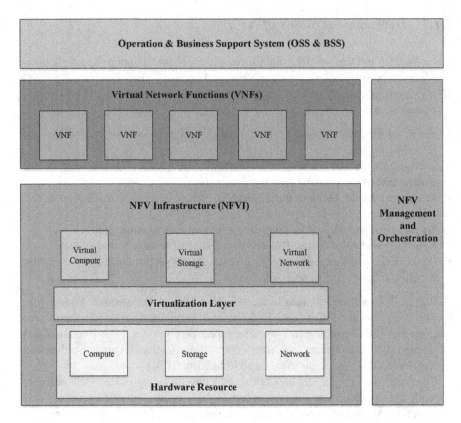

**Fig. 2.2** NFV architecture

- **Services network** is the telecommunication service networks.
- **Management and orchestration** is the most significant difference between NFV and traditional network, referred to as MANO. MANO is responsible for the management and orchestration of the overall NFVI resources, business network and mapping and association of NFVI resources, and the implementation of OSS business resource process.

According to the NFV technology principle, a business network can be decomposed into a set of VNF and VNF Link (VNFL), represented as VNF Forwarding Graph (VNF-FG). Each VNF consists of several VNF Components (VNFC) and an internal connection diagram, and each VNFC is mapped to a Virtual Machine (VM). Each VNFL corresponds to an Internet Protocol (IP) connection, which needs link resources, such as flow, Quality of Service (QoS), routing, and other parameters. Thus, the services network can make top-down dissolutions to get distributable resources through MANO. The corresponding VM resources and other resources are allocated by NFVI. In addition, the corresponding VNFL resources need to interact

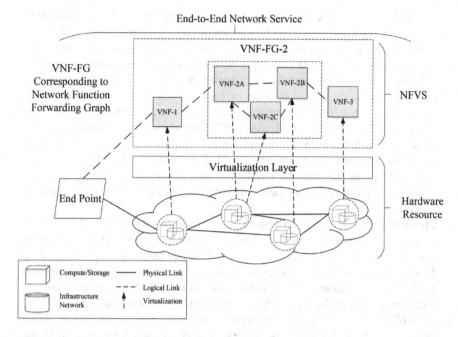

**Fig. 2.3** Services network deploying NFV

with the bearer network management system, and to be allocated by IP bearer network. For example, Fig. 2.3 illustrates the services network deploying NFV.

According to the current technical architecture of NFV, many manufacturers have already completed the proof of concept (POC) testing and verification, such as virtual IP Multimedia Subsystem (vIMS) [1], virtualized Evolved Packet Core (vEPC) [7], virtual Customer Premise Equipment (vCPE) [8], and virtual Content Distribution Network (vCDN) [3]. And they have been demonstrated at the annual meeting of the World Radio Communication Conference (WRC) in 2014 to prove that NFV technology is available.

## 2.1.2 Technical Issues of NFV

Although the criterion defined by NFC is technically feasible, there is still a long way to realize its commercial application with the following issues [4–6]:

- **Maturity**: Due to its too large target, only four specifications have been completed after the first phase, while many relevant specifications defined by other groups estimate to complete. Many problems have been postponed to the second phase, so there is still a long way to go to meet its mature standard.

- **Compatibility**: Architecture defined by NFV is quite huge with many new inter-faces, dividing the closed telecom equipment manufacturers into several levels: hardware equipment suppliers, virtualization management software suppliers, virtualization software vendors, NFV Orchestrator (NFVO) software vendors, NFV system integrator, etc. Thus, the telecom network is transferred from a integration of hardware and software managed by one manufacture into a series integrations of hardware and software managed by several manufactures, so the complexity increases greatly. However, NFV only defines the architecture levels, while the detailed definition and implementation of the corresponding interfaces are to be coordinated by other technical organizations. Therefore, compared with the existing standard, the technical standards are not so strict. It is a great challenge to ensure the equipment compatibility among various manufactures in the future.
- **Flexibility**: The lagging Self-Organization Network (SON) technology affects the expansion and deduction of service level. According to the NFV architecture, although the needed resources of a new VNF are automatically deployed by MANO, its business network operational architecture still relies on the traditional EMS/NMS mechanism, and the connection between VNF and traffic routing is still deployed manually and the VNF plug and play is not available.
- **Reliability**: Traditional telecom applications often require the reliability of 99.999 %, which should not be decreased after its virtualization. Due to the special design, the reliability requirements of traditional telecom hardware are relatively high. However, the reliability of COTS equipment adopted by the virtualization is relatively lower, demanding compensation by raising the software reliability.
- **Integration**: The current telecommunications equipment often uses special chips to realize user plane. Considering the packet mangling, x86 has lower cost performance. Therefore, its virtualization will lead to the reduction of equipment integration. Currently there are several ways to solve this problem: (1) the Software Defined Network (SDN) is implemented to separate the control and operation of user plane equipment and offload the forwarded packet to the SDN switch; (2) the Intelligent Ethernet Card including packet processing module is implemented to offload packet processing burden.
- **Virtualization**: Compared with computing and storage virtualization, network virtualization technology is relatively backward. Although the current network virtualization technology has various types, it is a critical issue to integrate them into the NFVI. Telecommunication network is usually a distributed network needing sufficient network resources, which are decomposed to local network resource within data center, the bearer network resources among the data center, the bearer network resources between the service network and access network, etc. The allocation of the bearer network resources may involve the transport network resources allocation, which needs virtualization and automa-tion. Currently the allocation still needs to fulfil through bearer network and transport network management, which is a long way to reach the automation.

- **Systematicity**: NFV is expected to solve the problem of automatic deployment of business network, which is a giant Information and Communication Technology (ICT) integration project from the perspective of architecture. NFV can be decomposed into NFVI integration, VNF integration, and business network integration, involving a number of systems, manufactures, areas, and interfaces, which makes the engineering more difficult than the current public/private cloud. Despite its automatic deployment, every link of the telecom network deployment (planning, implementation, testing, upgrade, optimization, operations, etc.) is involved and implemented. Therefore, it is a complicated issue to implement the deployment in the future, because the technical requirement for the integrator is very high.

After the implementation of NFV architecture, automatic management and agility of the telecom network should ascend dramatically. The deployment cycle of a telecommunications device is decreased from a few months to a few hours, the expansion cycle is decreased from a few weeks to a few minutes, and the new business deployment cycle of the telecommunications network is decreased from a few months to a few weeks.

## 2.2  Cloud Radio Access Networks

Cloud Radio Access Networks (Cloud-RAN) is a new type of wireless access network architecture based on the trend of current network conditions and technological progress. As a type of clean system, C-RAN is based on the Centralized Processing, Collaborative Radio, and Real-time Cloud Infrastructure. Its essence is to cut down the number of base station and reduce the energy consumption, adopt the collaboration and virtualization technology to realize the resources sharing and dynamic scheduling, improve the spectrum efficiency, and achieve low cost, high bandwidth, and flexible operation. C-RAN's overall goal is to address the various challenges brought by the rapid development of mobile networks, such as energy consumption, construction and operation and maintenance costs, and spectrum resources., pursuing a sustainable business and profit growth in the future [2].

As shown in Fig. 2.4, C-RAN architecture mainly consists of the following three components:

- Distributed network consisting of a Remote Radio Unit (RRU) and an antenna.
- Optical transmission network with high bandwidth and low latency which connects the RRU and the Bandwidth-Based Unit (BBU).
- Centralized base band processing pool consisting of high performance general processor and real-time virtual technology.

C-RAN architecture includes the following advantages:

- The centralized approach can greatly reduce the number of base stations and the energy consumption of the air conditioning systems.

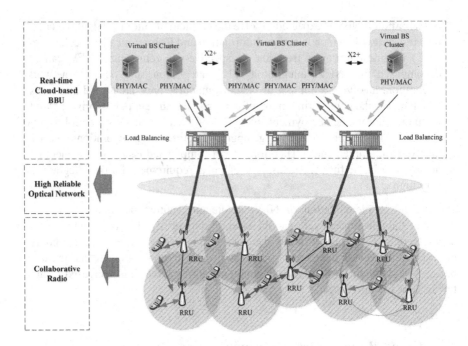

**Fig. 2.4** C-RAN architecture

- Due to the high-density RRU, the distance from RRU to the users is shortened
  for reducing the emission power without affecting the overall network coverage.
  Low transmission power means that the terminal's battery life will be longer and
  the power consumption of wireless access networks will be reduced.

Different from the traditional distributed base station, C-RAN breaks the fixed
connection relationship between RRU and BBU that each RRU does not belong
to any BBU. Sending and receiving signals in RRU is in a virtual BBU, while
the processing capacity of the virtual base station is supported by the assigned
processors in the real-time virtual allocation base band pool.

In the C-RAN architecture, the sites of BBU can be reduced by one to two orders
of magnitude. Centralized base band pool and related auxiliary equipment can be
placed in some key central machine room for simple operation and management.
Though the number of RRU is not reduced in C-RAN, due to the small size and low
power consumption of these devices, they can be easily deployed in a limited space
with the power supply system and without the frequent maintenance. As a result, it
can accelerate the speed of the operational network construction.

## 2.3 Mobile Cloud Networking

Mobile cloud networking (MCN)[2] is a large-scale integrated project funded by the European Commission EP7, focusing on the implementation of cloud computing and network function virtualization to achieve the virtual cellular network. It is designed as a completely cloud-based mobile communication and application platform. More specifically, it aims to investigate, implement, and evaluate the LTE mobile communication system's technology base. This mobile communication system provides atomic level of service based on the mobile network and decentralized computing and intelligent storage, in order to support atomical services and flexible payment.

As shown in Fig. 2.5, MCN is expected to achieve the following goals:

- MCN is expected to provide the basic network infrastructure and platform software as a service for solving the resources waste problems (energy, bandwidth, etc.) facing the inflexible traditional network, and supporting payment on demand, self-service, flexible consumption, remote access, and other services.
- The structure of cloud computing is unable to support the integration with the mobile ecosystem. Therefore, MCN attempts to extend the cloud computing concept from data center to the mobile terminal users. Specifically, the new virtualization layer and monitoring system is designed, the new mobile platform is developed for the future mobile services and application supporting cloud, and the end-to-end MCN services are provided.

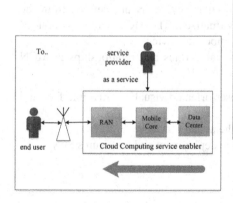
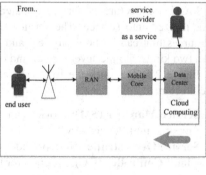

**Fig. 2.5** The goals of MCN

---

[2]http://www.mobile-cloud-networking.eu/site/.

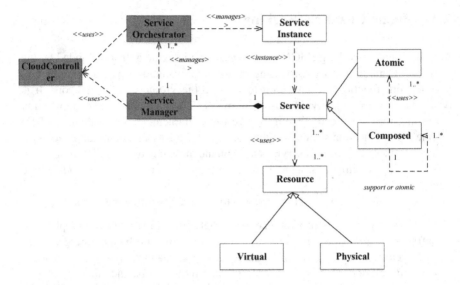

**Fig. 2.6** The crucial entities and relationships in MCN architecture

MCN focuses on two main principles: (1) the cloud computing service must illustrate the resource pool, (2) the architecture is service-oriented. The related work of MCN mainly consists of the following components: cloud computing infrastructure, wireless cloud, mobile core network cloud, and mobile platform services.

MCN architecture is service-oriented, in which the functional elements are modularized into service. The services provided by MCN are derived from the resources that can be both physical and virtualized. The MCN service is divided into two kinds: atomic-level service and composite service.

Figure 2.6 illustrates the following crucial entities and relationships in MCN architecture:

- **Service Manager (SM)**: provides an user-oriented visual external interface and supports multi-tenant services.
- **Service Orchestrator (SO)**: provides the actual services.
- **Cloud Controller (CC)**: supports for the deployment and configures SOs.

## References

1. G. Carella, M. Corici, P. Crosta, P. Comi, T.M. Bohnert, A.A. Corici, D. Vingarzan, T. Magedanz, Cloudified IP multimedia subsystem (IMS) for network function virtualization (NFV)-based architectures, in *2014 IEEE Symposium on Computers and Communications (ISCC)* (IEEE, Madeira, 2014), pp. 1–6

2. A. Checko, H.L. Christiansen, Y. Yan, L. Scolari, G. Kardaras, M.S. Berger, L. Dittmann, Cloud ran for mobile networks—a technology overview. IEEE Commun. Surv. Tutorials **17**(1), 405–426 (2015)
3. K. Heo, I. Jung, C. Yoon, A study of enhancement in virtual content distribution network, in *2014 International Conference on Information Science & Applications (ICISA)* (IEEE, Seoul, 2014), pp. 1–5
4. Z. Huiling, S. Fan, Development and challenge of SDN/NFV. Telecommun. Sci. **30**(8), 13–18 (2014)
5. M. Jarschel, A. Basta, W. Kellerer, M. Hoffmann, SDN and NFV in the mobile core: approaches and challenges. IT-Inf. Technol. **57**(5), 305–313 (2015)
6. J. Liu, Z. Jiang, N. Kato, O. Akashi, A. Takahara, Reliability evaluation for NFV deployment of future mobile broadband networks. IEEE Wirel. Commun. **23**(3), 90–96 (2016)
7. M.R. Sama, L.M. Contreras, J. Kaippallimalil, I. Akiyoshi, H. Qian, H. Ni, Software-defined control of the virtualized mobile packet core. IEEE Commun. Mag. **53**(2), 107–115 (2015)
8. K. Suksomboon, M. Fukushima, M. Hayashi, Optimal virtualization of functionality for customer premise equipment, in *2015 IEEE International Conference on Communications (ICC)* (IEEE, London, 2015), pp. 5685–5690

# Chapter 3
# Cloud Platform for Networking

**Abstract** It is known that cloud computing is a kind of Internet-based computing that provides shared processing resources and data to computers and other devices on demand. It is a model for enabling ubiquitous, on-demand access to a shared pool of configurable computing resources. With the development of NFV, SDN and other advanced networking technologies, cloud platform is widely used to manage virtual network resources and functions for providing more connectivity choices, better performance, and lower prices.

## 3.1 OpenNebule

OpenNebula[1] is a open source toolbox for cloud computing, and its overall architecture as shown in Fig. 3.1. It supports to establish and manage the private cloud with the implementation of Xen [1], Kernel-based Virtual Machine (KVM) [4], or VMware ESX [6], and provides Deltacloud[2] adapter collaborative with Amazon Elastic Compute Cloud (Amazon EC2) [10] to manage the hybrid cloud. Besides the cloud service providers like Amazon, the Amazon partners running the private cloud on the different OpenNebula instances can also play the role of the remote cloud service providers. The current version of OpenNebula supports XEN, KVM, and VMware, as well as real-time access to EC2 and ElasticHosts.[3] Furthermore, it supports the image file copy and transmission, and virtual network management.

OpenNebula provides the following functions to the enterprise for implementing the private cloud, hybrid cloud, and public cloud:

1. Highly secure multi-tenant operations;
2. On-demand preparation and monitoring of computing, storage, and network resources;
3. High availability;
4. Distributed resource optimization to provide better workload performance;

---

[1] http://opennebula.org/.

[2] https://deltacloud.apache.org/.

[3] https://www.elastichosts.com/.

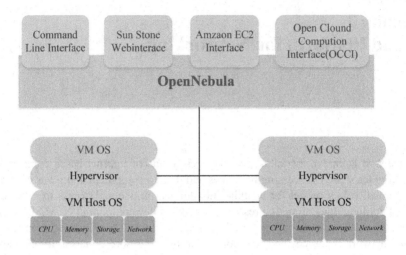

**Fig. 3.1** OpenNebula architecture

5. Centralized management across multiple regions and available interfaces;
6. High scalability.

As shown in Fig. 3.2, the private cloud aims to give local users and administrators with a flexible and agile private infrastructure, to run the virtual services in the manageable domain managed. OpenNebula virtual infrastructure exposes the Application Programming Interfaces (APIs) of virtualization, networking, image and physical resources configuration, management, monitoring, and accounting. An OpenNebula private cloud provides the users with a fast delivery and scalable infrastructure platform to meet the dynamic demands. The services are hosted in a virtual machine, and then submitted, monitored, and controlled through OpenNebula operations center or OpenNebula interfaces in the cloud.

As shown in Fig. 3.3, OpenNebula provides Deltacloud adapter and Amazon EC2 to manage the hybrid cloud.

The OpenNebula public cloud is an extension of the private cloud to expose the Representational State Transfer (REST) interface. If you permit your partners or external users access to your infrastructure or to sell your services, the cloud interface should be added to your private or hybrid cloud. Obviously, a local cloud solution is the natural backend for any public cloud.

As shown in Fig. 3.4, OpenNebula framework consists of three layers: the drivers layer, the core layer, and the tools layer. The drivers layer directly interacts with the operating system to create, startup, and shut down the VMs, allocate storage for the VMs, and monitor the status of the physical and virtual machines. The core layer manages the VMs, storage devices, and virtual networks. The tools layer provides the users with the APIs and the command line or browser as the user interface.

OpenNebula uses the shared storage devices to provide VM images so that each compute node can access the same VM image resource. If users need to start or shut

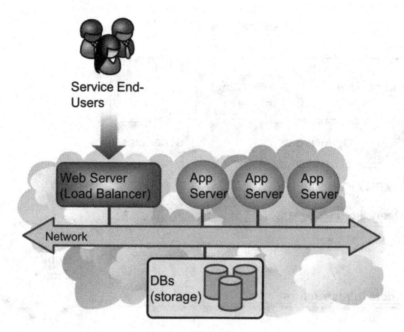

**Fig. 3.2** OpenNebula private cloud (*Source*: OpenNebula project)

down a VM, OpenNebula will login to the compute node to directly running the corresponding virtualization management commands. This model is also known as agentless [7] to eliminate the need to install the additional software (or service) on the compute nodes, so the system complexity is relatively lower.

Moreover, OpenNebula uses the bridge to connect the virtual network, while the IP and Media Access Control (MAC) address of each node is generated within a certain range. The network will be connected to a specific bridge, while each bridge has his network owner and it can be public or private. The virtual network is isolated from each other, and it uses Ebtables[4] to filter the data link layer packet.

## 3.2   OpenStack

OpenStack[5] is a cloud operating system for managing data center computing, storage, networking, etc., which can be used to create public and private cloud [8]. It is expected to establish an open standard for cloud computing platform to provide the companies with the solution of infrastructure as a service (IAAS). Currently,

---

[4]http://ebtables.netfilter.org/.
[5]https://www.openstack.org/.

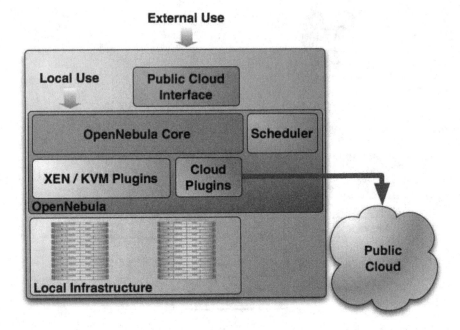

**Fig. 3.3** OpenNebula hybrid cloud (*Source*: OpenNebula project)

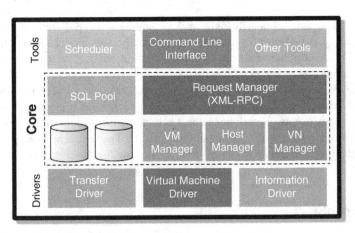

**Fig. 3.4** OpenNebula tri-layer architecture (*Source*: CloudUser, 2010)

hundreds of organizations contribute to its source code, and the open source community is completely transparent management, design, and development, and the underlying provide the upper application with computing, storage, and network resources through open API. OpenStack is mainly programmed by Python,[6] and its architecture is designed with a completely decoupled modularized ideas. Therefore, OpenStack has a very good openness and compatibility.

OpenStack consists of the following five components:

1. Keystone provides authentication service.
2. Nova provides computing service.
3. Swift provides storage service.
4. Glance provides image service.
5. Horizon provides dashboard service.

Especially, Horizon is a Python-based Web framework developed by Django[7] for visually managing OpenStack platform. Nova is the computing controller of OpenStack that it allocates on-demand VM according to the user's requests and manages the virtual computing resource allocation and scheduling, which is the component for manage the allocation and scheduling of VM. In OpenStack, Nova processes the scheduling from VM creation to termination of the life cycle. Actually, the VM isn't operated by Nova directly, but processed by the underlying operating system Hypervisors through libvirt[8] API [3].

As shown in Fig. 3.5, Nova consisting of the following modules, provides the user with APIs to operate and manage VMs, while the cloud infrastructure must be managed through Nova-API.

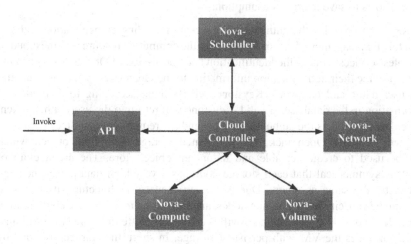

**Fig. 3.5** NOVA

---

[6]https://www.python.org/.
[7]https://www.djangoproject.com/.
[8]http://libvirt.org/.

- **Message Queue** is the communication module between each node in OpenStack, which is mainly based on Advanced Message Queue Protocol (AMQP). Since major operations of Nova are very time-consuming, in order to reduce the user response time, Nova responses the use's request asynchronously through callbacks.
- **Nova-Compute** is used to manage the life cycle of the instance, which is often a VM. After receiving the request to create or terminate a VM, Nova-Compute process it through libvirt API and then return the results by the message queue.
- **Nova-Network** provides the VM with network connection services that the intra and inter network communications of the VMs are processed by this module. Specifically, Nova-Network is mainly in charge of assigning IP address to the VM, Virtual Local Area Network (VLAN) and security groups configuration, etc.
- **Nova-Volume** provides the VM with persistent storage, which is a very important for the computing devices and can greatly reduce the losses caused by power outages, downtime, and system-level failure.
- **Nova-Scheduler** is a daemon starting at the initiation of the cloud platform. When Nova receives the request to create a VM, Nova-Scheduler needs to decide which computing node should be used to create the VM. When the VM needs to be migrated, Nova-Scheduler manages the VM migration and resource redistribution. The VM migration is a very complex process that Nova-Scheduler needs to avoid wasting computing resources and ensure the cloud platform overall performance is not decreased during the VM migration, i.e., sleeping the idle hosts to save energy consumption.

Keystone provides the authentication service including authentication and service token management. User can't access to the computing resources in the cloud or operates services without the identification and permission of OpenStack. When the users provide their authentication information to the OpenStack, which is generally the user name and password, Keystone verifies them according to their identity information in the database. If valid, Keystone will return to the user with a Token, which can be used as the authentication to send the request to OpenStack.

Swift provides OpenStack with distributed storage for virtual object, which can be used to create scalable and redundant object store. The architecture of Swift is symmetrical that each storage node has a very high data persistence and is exactly the same as others. Due to the symmetrical architecture, it is easy to expand the capacity just by adding nodes, and there is no master–slave configuration dependence or single node failure. Swift is completely different from Nova-Volume, which provides the VM with persistent storage. In short, the storage provided by Nova-Volume is similar to the hard disk, while Swift based on a distributed approach mainly supports massive object storage and provides the VM and cloud applications with data containers, secure storage, data backup, etc.

Glance is used to store and retrieve the VM image. When OpenStack creates a VM, it is available to retrieve the VM image by Glance and regenerate the original VM via the copy or snapshot of its image. Furthermore, Glance provides the standard REST interface to query the image information stored on the different devices.

## 3.3 OpenDayLight

OpenDaylight[9] is a community-driven open source framework to promote the innovation and implementation of SDN. Faced SDN, the right tools are essential to manage the infrastructure, which is the expertise of OpenDaylight. OpenDaylight has a modular, pluggable, and extremely flexible controller including a collection of modules can quickly complete the network tasks, which enable it to be deployed on any Java-enabled platform [5].

In Fig. 3.6, it illustrates that the architecture of the latest OpenDaylight consisting of four layers. OpenDaylight provides the applications with the opened northbound API and supports Open Services Gateway initiative (OSGi) framework and bi-directional REST API. Specifically, the OSGi framework is provided to the applications running in the same address with the controller, while the REST API is provided to the applications running in the different address.

The control platform includes the basic network services and some additional services installed as a plugin, which increases the flexibility of OpenDaylight. Of course, it is stable, but it is not as stable as Open Network Operating System (ONOS) which is a distributed strategy.

The southbound supports various protocols through the plugins, including OpenFlow, Border Gateway Protocol Link-State (BGP-LS), etc. These modules are dynamically mounted to the service abstraction layer (SAL) for the upper service that the call from the upper layer packaged as a suitable protocol format for the underlying network devices. However, one of the southbound protocol

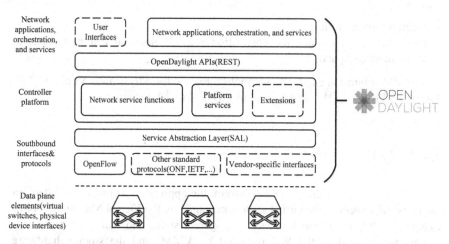

**Fig. 3.6** OpenDayLight framework (*Source*: OpenDaylight: an open source SDN for your OpenStack cloud, OpenDaylight, 2013)

[9]https://www.opendaylight.org/.

named OpFlex [9] is questionable, which is considered by some to be a wrong abstraction and expose the details of the device to the application, which means that it involves less abstraction and more complexity. It can be seen that the southbound of OpenDaylight does not completely abstract the underlying devices to be processed by the controller platform.

## 3.4   Virtual Machine Migration

The migration is able to save money for management, maintenance, and upgrade, and enables a single server to simultaneously replace the previous multiple servers for saving a lot of room space. In addition, the VM has a unified "virtual hardware resources" unlike the previous server has various different hardware resources, such as the different chipset, network cards, hard drives, and Graphics Processing Unit (GPU). After the migration, the VMs can be managed in a unified interface, and one VM can be switched to another through some VM software when it goes down, which supports the uninterrupted service. In short, the migration has the advantage of simplifying the system maintenance and management, improving the system load balancing, enhancing the system error tolerance, and optimizing the system power management.

The excellent migration tool is expected to minimize the overall migration time and downtime, and limit the negatively performance impact caused by the migration. Specifically, the VM migration performance indicators include the following three aspects:

1. Overall migration time;
2. Downtime, i.e., the source and destination hosts are unavailable at the same time;
3. Performance impact.

The VM migration can be divided into three modes: Physical-to-Virtual (P2V), Virtual-to-Virtual (V2V), and Virtual-to-Physical (V2P) [2].

### 3.4.1   P2V

P2V refers to migrate the operating system, the application software, and data on the physical servers to the virtual servers managed by the Virtual Machine Monitor (VMM). In this migration approach, the system status and data on the physical servers are imaged to the VM provided by VMM, and the storage hardware and network adapter driver are replaced in the VM. With the installation of the appropriate drivers and configuration of the same address as the original server, the VM is available to work as the original physical machine alternative after restarting.

P2V migration method is divided into manual migration, semi-automatic migration, and live migration.

- **Manual Migration**: The migration is manually completed based on the fully understanding about the system environments of the physical machine and the VM. Firstly, the service and the operating system on the original physical machine are shut down, and a new system is booted in other media. Secondly, the disk of the physical machine is imaged as a VM image file. Especially, if more than one disks in the physical machine, all the disks will imaged to the VM. Thirdly, the virtual devices are created and the image files are loaded for the VM. Finally, the VM is started to adjust the system settings and turn on the services.
- **Semi-automatic Migration**: Assisted by the professional tools, such as Virt-P2V and[10] Microsoft Virtual Server Migration Toolkit,[11] some manual operations of P2V migration are automatized. For example, the data format conversion is often a very time-consuming task, but it is convenient to finalized it through some professional tools.
- **Live Migration**: Most P2V tools have a great limitation that the physical machine is unavailable during the migration. Fortunately, with the development of P2V technology, VMware vCenter Converter[12] and Microsoft Hyper-V[13] have been able to provide live migration and avoid downtime. Currently, only Windows supports P2V live migration.

## 3.4.2   V2V

V2V is the operating system and data migration between the VSs, considering the machine-level differences and dealing with the different virtual hardware. The VM migrates from a physical machine to another, while the VMM on these two physical machine may be the same or different, such as the migration from VMware to KVM or from KVM to KVM. There are several ways to migrate the VM from one VM Host system to another. V2V migration can be divided into offline (also known as the static migration) and online migration (also known as live migration).

Before the offline migration, the VMs are paused. If the VM uses the shared storage, the migration will be simple that only the system status is copied to the destination host for recreating this VM. If the VM uses the local store, not only the system status but also the image of the VM should be copied. The offline migration is relatively simple, but there is a serious drawback that the VM must be stopped providing services.

---

[10]http://libguestfs.org/virt-p2v.1.html.

[11]https://www.microsoft.com/en-hk/download/details.aspx?id=31022.

[12]http://www.vmware.com/products/converter.html.

[13]https://www.microsoft.com/en-us/cloud-platform/virtualization.

The online migration overcome the shortcomings that the VM must be stopped during the offline migration, so the VM is available during the migration. Actually, the original machine is also stopped, but there is a short pause for switch the VM while the destination host is available to provide the services. Since the switching is very rapid, the migration does not affect the services provided by the VM. The online migration is also divided into the shared storage model and the local storage model. With the local storage, the system status, memory data, and disk image are migrated via block migration in OpenStack. In the KVM, at the beginning of the block migration, the system disk and data disk are created on the destination host member with the same path on the source host, and a VM with the same configuration as the original VM is created on the destination host through Libvirt API. Then the data migration is started, while the source VM is still running, so the data on these two VM should be synchronized after the migration. Finally, the source VM is shut down while the destination VM starts providing the services.

### 3.4.3   V2P

V2P is the inverse operation of P2V that the operating system, application, and data are migrated from a VM to a physical machine. In particular, a VM can be migrated to one or more physical machines, though the virtualization is expected to transform the physical machine as VM. For example, some bugs of the applications on the VM must be verified on the physical machine. Additionally, it is difficult to configure a new workstation, but the virtualized application can help to solve this problem via P2V. However, this approach has two limitations: the image must be mounted on the same hardware infrastructure, and each changed configuration is saved by renew the image. V2P migration can be completed manually, but it is better to simplify the operation assisted by the migration tools, such as PlateSpin Migrate,[14] especially in the case of involving various different hardware.

### References

1. P. Barham, B. Dragovic, K. Fraser, S. Hand, T. Harris, A. Ho, R. Neugebauer, I. Pratt, A. Warfield, Xen and the art of virtualization, in *Proceedings of the Nineteenth ACM Symposium on Operating Systems Principles, SOSP '03* (ACM, New York, NY, 2003), pp. 164–177. ISBN 1-58113-757-5. doi:10.1145/945445.945462. http://doi.acm.org/10.1145/945445.945462
2. D. Barrett, G. Kipper, *Virtualization and Forensics: A Digital Forensic Investigator's Guide to Virtual Environments* (Syngress, Burlington, 2010)
3. M. Bolte, M. Sievers, G. Birkenheuer, O. Niehörster, A. Brinkmann, Non-intrusive virtualization management using libvirt, in *Proceedings of the Conference on Design, Automation and Test in Europe* (European Design and Automation Association, Dresden, 2010), pp. 574–579

---

[14]https://www.netiq.com/products/migrate/.

4. A. Kivity, Y. Kamay, D. Laor, U. Lublin, A. Liguori, KVM: the Linux virtual machine monitor, in *Proceedings of the Linux Symposium*, vol. 1 (2007), pp. 225–230
5. J. Medved, R. Varga, A. Tkacik, K. Gray, Opendaylight: towards a model-driven SDN controller architecture, in *Proceeding of IEEE International Symposium on a World of Wireless, Mobile and Multimedia Networks 2014* (2014)
6. A. Muller, S. Wilson, *Virtualization with VMware ESX Server* (2005)
7. M. Rose, F.W. Broussard, Agentless application virtualization: enabling the evolution of the desktop. White Paper, IDC, and Sponsored by VMware (2008)
8. O. Sefraoui, M. Aissaoui, M. Eleuldj, Openstack: toward an open source solution for cloud computing. Int. J. Comput. Appl. **55**(3), 38–42 (2012)
9. M. Smith, M. Dvorkin, Y. Laribi, V. Pandey, P. Garg, N. Weidenbacher, OpFlex control protocol. IETF (2014)
10. G. Wang, T.E. Ng, The impact of virtualization on network performance of amazon EC2 data center, in *INFOCOM, 2010 Proceedings IEEE* (IEEE, San Diego, 2010), pp. 1–9

# Chapter 4
# Definable Networking

**Abstract** For the purpose of high reliability and short delay of the end-to-end, it is necessary to optimize and evolve the existing network architecture and the functional configuration of network element. SDN and NFV provide new thought for the systemic design of 5G, i.e., the software-driven flexible system architecture integrating with the infrastructure closely. Compared with the traditional IP network based on the distributed routing calculation, the SDN based on centralized routing calculation can effectively schedule all the network resources. NFV is available to connect all the network nodes and realize network intelligence by software programming for improving the flexibility. Furthermore, NFV enables the operators to meet the users' demands by controlling the network capacity for improving the scalability. The combination of SDN and NFV, the upgrade of the network application, and network hardware are separated. This chapter focuses on the definable networking, one of the key techniques supporting 5G network architecture.

## 4.1 Caching

Nowadays, caching technology is no longer a new technology to reduce the response time to the user's request through various ways, such as proxy caching [32] and transparent proxy caching [2]. Through Web caching, the network traffic is minimized when the user is surfing the Internet, while the access speed is also extensively raised [5]. As the core of Content Distribution Network (CDN) is to increase the access speed of the Internet, caching technology is one of the most important technology for CDN, and the Web cache server is the core of caching system and CDN [28].

Cache server can be a common server installing with the caching software or a special equipment including the software and hardware system. In brief, cache server plays the role of filtration and agency between the original server and users. It can filter the users redundant request to save the bandwidth of backbone network and increase the response speed. In addition, it can get the information for users from the original server as an agency. As shown in Fig. 4.1, it illustrates the position

© The Author(s) 2016
Y. Zhang, M. Chen, *Cloud Based 5G Wireless Networks*, SpringerBriefs
in Computer Science, DOI 10.1007/978-3-319-47343-7_4

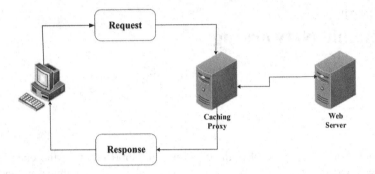

**Fig. 4.1** Caching server

and function of the cache server. If the requested content is reserved in the cache server, it will response to the user's request directly. Otherwise, the cache server will send the request to the original server on behalf of the user and then copy and retransmit the response to the user.

Web caching reserves the frequently accessed content in the cache server at the edge of network and provides the users with these contents rapidly. Thus, it could release the burden of the original server and the redundant flow of backbone network, while the response time is shortened.

## 4.2  Mobile Content Distribution Network

Mobile Content Distribution Network (MCDN) is expected to deliver the content to the terminal users in an optimal manner through any kind of wireless network or mobile network [9]. As same as the traditional CDN, MCDN aims to provide the terminal users with the content ensuring the high availability and high performance. Furthermore, MCDN can optimize the content delivering to the mobile devices through some particular wireless network, such as the limited network capacity and lower resolution. By enhancing the detection around the intelligent equipment, MCDN can solve the instinct challenges of the mobile network including the long delay, high packet loss rate, and large downloads [42].

MCDN is able to relief the traffic pressure of the core network by deploying CDN service nodes in the core mobile network or the lineside of wireless network. In the implementation of MCDN for the mobile network, the point is how to get the users' IP data grouping for analysis and respond in advance. Therefore, it is possible to deploy the cache equipment at NE for MCDN, such as Evolved Node B (eNode B), Serving GPRS Support Node (SGSN), and Gateway GPRS Support Node (GGSN) in the core network, as shown in Fig. 4.2.

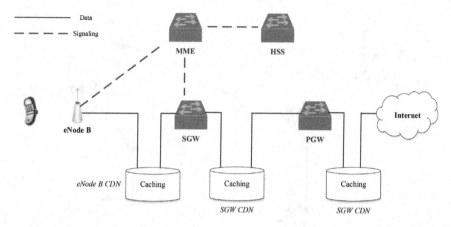

**Fig. 4.2** eNobe B MCDN

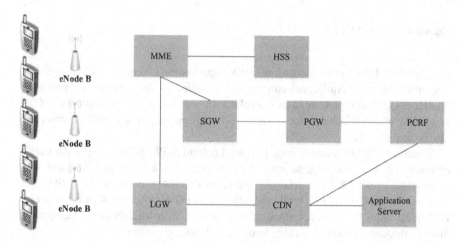

**Fig. 4.3** LGW-based MCDN

It is not appropriate to place MCDN storage node behind each eNode B, considering to the large number and wide distribution of eNode B. For the specific area containing a great many users, it is available to add Local Gateway (LGW) behind eNode B as a regional equipment, and the CDN is collaborative deployed behind LGW, as shown in Fig. 4.3. The existing network architecture keeps the same that the Packet Data Network Gateway (PGW) still serves as the provincial anchor equipment and the Mobility Management Entity (MME)/Serving Gateway (SGW) serves as the municipal anchor equipment. The function of the added LGW is the combination of SGW and PGW. Similar to SGW, LGW is under the control of MME, and LGW can include one or several eNode B. In the 3GPP draft, the LGW has been defined to realize the sideway CDN, but it hasn't become a common accepted standard. This solution just needs a little change in the existing network,

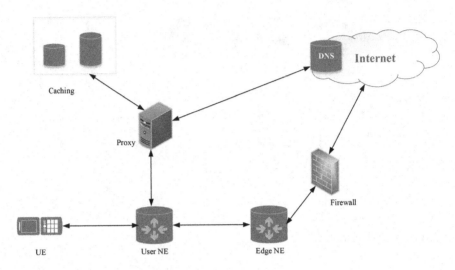

**Fig. 4.4** SGW-based MCDN

because the MME in the existing network supports the essential LGW and SGW.
Therefore, the new equipment can smoothly deploy in the network without new
interface. LGW can be treated as a weakened version of the combination of SGW
and PGW, which is suitable for some specific regions such as the busy commercial
places and schools.

When the MCDN storage node is placed behind SGW, SGW can get the user's
complete IP data grouping. Moreover, SGW is always a municipal NE, which is
appropriate to set MCDN cache. As shown in Fig. 4.4, the user's NE is SGW in
Evolved Packet Core (EPC) while the edge NE is PGW. After SGW gets the user's
IP request from the protocol stack, it can connect with MCDN cache equipment
directly through Ethernet interface to transfer IP data grouping.

When the MCDN storage node is placed behind PGW, because the position
of storage is near the coboundary of EPC, the traffic burden of core network is
decreased a little after placing the cache node. In some degree, it is just like a usual
CDN system.

## 4.3  Software-Defined Mobile Network

With the rapid development of the networking, it brings the great challenges to the
traditional Internet, such as the high complexity of network configuration, which
indicates that the network infrastructure needs to be innovated. Fortunately, the
researches on the programmable network provide the theoretical foundation for
SND. Active network supports the data packet carrying with user program, which
can be performed automatically by the network device. Therefore, it is available to

dynamically configure the network programmatically for the convenient network management. However, due to the low demand and incompatible protocol, it is not actually deployed in the industry. 4D approach, including four planes: decision, dissemination, discovery, and data, to network control and management is proposed to separated the decision planes (i.e., control planes) from the data plane, i.e., centralize and automatize the control plane [13]. The design ideas produce the prototype of SDN controller.

SDN is originated in 2006 from Stanford Clean Slate research. In 2008, Professor Mckeown et al. proposed the concept of SDN to separate the control and data through the hierarchical design [23]. In the control layer, including the central logic and programmable controllers, the global network information is provided to the operators and researchers for the new protocols deployment and the network management and configuration. In the data layer, the dumb switch only provides the simple data forwarding function, which is different from the traditional switcher, and the corresponding packets are processed quickly for meeting the traffic growing demand. The control layer interacts with data layer through the unified open interfaces, such as OpenFlow. Specifically, the controller sends the unified standard rules to the switch through the standard interfaces, and the corresponding tasks are processed by the switches following the rules. Therefore, SDN technology can effectively reduce the equipment load to help network operators to better control infrastructure, reduce overall operating costs. SDN has become one of the most promising network technologies. Therefore, SDN related researches develop rapidly in recent years.

Draw upon the abstract architecture of the computer system architecture, the future network architecture includes virtualization concepts, i.e., abstract forwarding, abstract distribution, and abstract configuration. The abstract forwarding removes the control functions from the traditional switches, while all the controls are completed by the control layer. Moreover, the standard interfaces are provided to ensure that the switch is able to identify and forward the data. The control layer needs to abstract the equipment distribution into the overall network view, so that the applications can be uniformly configured through the overall network information. The abstract configuration further simplifies the network model that the users only need to make some simple network configuration through the interfaces provided by the control layer, and then the unified deployment is automatically forwarded to the devices along the path. Therefore, the ideas of network abstraction decouple the dependence on the path, and become the determining factor to establish the architecture with the unified interface architecture and separating control and data, i.e., SDN.

In addition, many organizations have participated in the SDN-related standardization. Specifically, the Open Networking Foundation (ONF)[1] is a representative organization to standardize the interface of SDN. The OpenFlow protocol developed by ONF has become the mainstream standard of SDN interface that many operators

---

[1]https://www.opennetworking.org/index.php.

and manufacturers research and develop SDN product in accordance with this standard. The Forwarding and Control Element Separation (ForCES)[2] of Internet Engineering Task Force (IETF), Software-Defined Networking Research Group (SDNRG)[3] of Internet Research Task Force (IRTF) and ITU Telecommunication Standardization Sector (ITU-T)[4], and other working groups focus on the novel method and applications related to SDN. The standardization organizations promote the rapid development of the SDN market, and SDN has a broad development prospects and great research value.

### 4.3.1   SDN Architecture

Various SDN architectures have been proposed for addressing the different demands. The first SDN architecture is proposed by ONF, which is widely accepted in the academia and industry. In addition, the NFV architecture is proposed by ETSI for carrier networks, which is supported by the industry. OpenDaylight is jointly proposed by the major equipment manufacturers and software companies to concrete SDN architecture in the actual deployment.

The SDN architecture is originally mentioned by ONF, and Fig. 4.5 illustrates its version released in 2013. From bottom to top (or from south to north), the SDN architecture consists of data plane, control plane, and application plane. The communication between the data plane and control planes is supported by the SDN control-data-plane interface (CDPI), which is a unified communication standard mainly based on the OpenFlow protocol. The communication between the control plane and application plane is supported by the SDN northbound interfaces (NBIs), which is available to be customized according to the actual demands.

The data plane consists of the switches and other NEs, while the connections between the NEs follow the different rules. The control plane containing the SDN control logic is responsible for running the control logic, maintaining the network view. Furthermore, the controller abstracts the network view into network services, accesses CDPI proxy to invoke the appropriate network data path, and provides the operators, researchers, and other third parties with the convenient NBIs for customized private applications and logic network management. The application plane includes various SDN-based network applications. The users do not care about the technical details of the underlying device, and the new applications can be quickly deployed through simply programming. CDPI is responsible for forwarding the rules sent from the network operating system to the network devices, which needs to match the equipment with different manufacturers and models and does not affect the control layer and the upper logic. NBIs enable the third party to

---

[2]https://datatracker.ietf.org/wg/forces/documents/.

[3]https://irtf.org/sdnrg.

[4]http://www.itu.int/en/ITU-T/Pages/default.aspx.

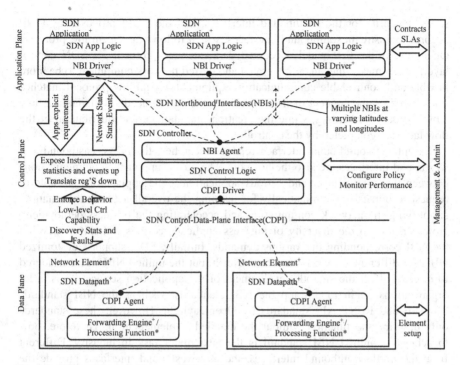

**Fig. 4.5** SDN architecture (*Source*: SDN Architecture Overview, ONF, 2013)

develop the individual network management software and applications, providing more options for managers. This network abstraction feature allows the user to select a different network operating system based on demand, and does not affect the normal operation of the physical device.

The interface in SDN is open, and the controller is logic center. The southbound is responsible for communications with data plane, while the northbound is responsible for the communications with application plane. Furthermore, because the single control mechanism is likely to cause control node failure, seriously affecting the performance, multi-mode controller is adopted.

In the researches on the open interfaces, the southbound interface of the controller is the core technology and research hotspot to separate control and data. Due to the decoupling between control layer and data layer, the improvements for these two layers are relatively independent that only the standard southbound interfaces are essential between the layers, which is a key component of the SDN layered architecture. Logically, it has to support the normal communication between the control layer and data layer, and the independent update of them. Physically, the manufacturers need to develop the devices supporting the standard southbound interfaces, because the traditional network equipment is not running among the SDN network. Therefore, the development of standard southbound interfaces is an important part of basic research on SDN. Many organizations have developed some standards for the southbound interfaces, and the CDPI proposed by ONF become mainstream southbound interfaces, which follows the OpenFlow protocol. OpenFlow is the first widely accepted data interface protocol for SDN control layer, which makes the single integrated and closed network equipment to be more flexible and controllable communication equipment. OpenFlow protocol matches rules based on the concept of flow, therefore, the switch needs to maintain a flow table for data forwarding, while the creation, maintained, and distribution of the flow table are completed by the controller.

Except the southbound interface related researches, the NBIs, eastbound and westbound interfaces are also the hotspots. The NBIs support the communications between the control layer and various applications. The applications are able to request resources programmatically for learning the overall network information, simplifying the network configuration and accelerating the business deployment. However, due to the diversity of business applications, it is difficult to develop the NBI corresponding the various demands. Initially, SDN supports customized NBIs for different scenarios custom fit North, but the unified NBIs are considered to directly affect the smooth development of the application service. Thus, a lot organizations begin to develop the NBI standards, such as the NBI standards by ONF and REST API standards by OpenDaylight. However, these standards only describe the function without the detailed implementation. Therefore, how to achieve a unified NBI standard is the next major task in industry. Different from the north-southbound interfaces, the east-westbound interfaces provide the communications between the controllers. Because the limited performance of the single controller cannot meet the demand for deploying the large-scale SDN, the east-westbound interfaces are expected to improve the salability of the controller and provide technical support for load balancing and performance improvements.

### 4.3.2  The Critical Techniques for Data Layer

In SDN, the data layer is separated from the control that the controller is responsible for the massive control strategy while the switch is only responsible for the fast packet forwarding according to the corresponding rules. In order to avoid frequent interaction between the switch and the controller, the rules are based on the flow rather than on a per-packet. The function of SDN data layer is relatively simple, and the related technology researches focus on: (1) switch design issues, i.e., how to design a scalable and fast forwarding device for flexibly matching rules and fast forwarding the data stream; (2) forwarding rules design issues, such as consistency update rules after failure problems.

**Switch Design Issues**

The SDN switch is deployed at the data layer for forwarding data flow by hardware and software. In general, the processing speed of the switch chip is 10–100 times faster than CPU, 1–10 times faster than the network processor (NP), and this difference will last a long time. However, in terms of flexibility, the switch is much lower than CPU, NP, and other programmable devices. Hence, it is the great challenge to design an advanced switch providing considerable forwarding rate and ensuring the flexibility to identify the forwarding rules.

The data processing by hardware can provide effective forwarding, but it is difficult to support flexible rule processing. In order to solve the strict rule matching and the limited actions on hardware, Bosshart et al. proposed RMT model to implement a reconfigurable matching table supporting the flow table with any width and depth in the pipeline stage [6]. A reconfigurable data layer should includes the following features.

1. The domain definition can be replaced or added.
2. The number of tables, topology, width, and depth can be specified. The limitation only is cased by the overall resources of the chip, such as memory chip size.
3. It is available to create new actions.
4. Packets can be placed into different queues, and the send port can be specified.

As another flexible processing technology through hardware, FlowAdapter switch is a layered approach for efficient and flexible multi-table line business [26]. The FlowAdapter switch consists of the following three layers:

1. The top layer is the software data plane supporting new protocols through updating.
2. The relatively fixed bottom layer is the hardware data plane for highly effective forwarding.
3. The middle layer provides the communications between the software data plane and the hardware data plane.

When the rules are distributed from the controller, the software data plane will store these rules as an M-phase flow table. Since these rules are relatively flexible, not all of them can be directly converted into the corresponding forwarding action by the switch, while the hardware data plane can provide high-speed matching and forwarding rules. So with the intermediate layer of FlowAdapter, the rules of these two data planes can be seamlessly transformed, i.e., the relatively flexible M-phase flow table is converted into an N-phase flow table being able to be identified by the hardware. For the purpose of conversion, the FlowAdapter checks all the rules of the software data plane, and then the M-phase flow table is converted into a 1-phase flow table according to the full rules. Finally, this 1-phase flow table is converted into an N-phase flow table sent to the hardware data plane. Through this seamless conversion, it theoretically solves the incompatibility of the multi-table pipeline technology between the traditional switch hardware and the controller. In addition, compared with the controller, the FlowAdapter is completely transparent, so updating FlowAdapter switch does not affect the normal operation of the controller.

Different from the hardware switch, though the processing speed of the software is lower than the hardware, the software can extensively improve the flexibility of the rules processing, and avoid the issues caused by the limited memory, the limited size of flow table size, unable processing the unexpected traffic, etc. By using the forwarding rules of the switch CPU, the problem of poor flexibility of hardware can be avoided. Because the CPU capability for processing packet has been improved increasingly and naturally deployed in the commercial switch. Thus, the processing speed difference between software and hardware is decreased, while the ability of flexible forwarding has been improved. Furthermore, it is available to process forwarding by NP. Since the NP is used to handle various networking tasks, such as packet forwarding, routing lookup, and protocol analysis, the processing ability of NP is more powerful than CPU. Regardless of using CPU or NP, it should take advantage of the flexible processing, and avoid the inefficient process.

In addition, in the data plane, it should be considered that the elements can be processed by hardware or software. For example, the original hardware design of the counter is not reasonable that the counter should be placed in the CPU to ensure the flexibility, saving the hardware space, reduce the complexity, and avoid the restrictions of hardware counter.

**Forwarding Rules Design Issues**

Similar to the traditional network, the network node failure problem is also possible in SDN, which results in the unexpected changes of the network forwarding rules seriously affecting the network reliability. In addition, the network traffic load transferring or network maintenance also brings about the changes in forwarding rules. SDN allows the administrators to freely update the rules through low-level abstraction management, which may result in the inconsistencies during updating the rules.

The inconsistencies caused by low-level management can be avoided by high-level abstraction management, which usually includes the following two procedure to updating the rules:

1. When a rule needs to be updated, the controller asks each switch whether they have finalized the flow corresponds to the old rules, and updates the rules of all the switches completing the processing.
2. When all the switches are updated, the updating is finalized, otherwise, it should be canceled.

For enabling the two-procedure rules updating, the data packet is preprocessed to be marked with a tag describing the rule version. During the forwarding, the switch checks the marked version, and processes the appropriate forwarding action corresponding to the related rules. After the packet is forwarded from the switch, then remove the tag. However, through this approach, the data packet corresponding to the new rules cannot be processed until all the data packet corresponding to the old rules have been processed, which results in the rules space to be occupied and higher costs. Fortunately, incremental consistency update algorithm can solve this problem that the rule updating is divided into several rounds, while each round is a two-procedure approach to update a subset of the rules, so the rules space can be saved.

Furthermore, McGeer proposed an OpenFlow-based updating protocol to improve the security of the two-procedure updating, which sends unrecognized packets to the controller for ensuring the accuracy of forwarding [22]. In addition, Ghorbani et al. proposed the rules updating algorithm for the VM, which ensures that the bandwidth is adequate during the updating while the normal forwarding is not affected [12].

Because the OpenFlow does not support to add new protocols, the protocol has to been updated into the interface specification for ensuring the new protocol can be used in the SDN. Hence, the protocol-independent forwarding is one of the most important researches on scalable data plane, especially the protocol-independent flow instruction set (FIS). The protocol-independent FIS abstracts the data plane that each rule related to the protocol is converted into a protocol-independent FIS and can be recognized by the data plane hardware for fast forwarding. The protocol-independent FIS separates the rules from the forwarding equipment to improve the data plane scalability and completely a full separation of control plane and data plane.

### 4.3.3   The Critical Techniques for Control Layer

The controller is the core component of the control layer. Through the controller, the user can logically and centrally control the switch for fast data forwarding, convenient and secure network management, and network performance improvement. As the first OpenFlow controller, the NOX[5] based improvement is discussed in this section, including two modes: the multi-threaded control mode and the flat and hierarchical control mode by increasing the number of distributed controllers. Furthermore the mainstream language interface for control language abstraction is introduced, while the in-depth analysis of the controller consistency, availability, and fault tolerance features are presented.

---

[5]https://github.com/noxrepo/nox.

## Controller Design Issues

The basic function of the controller is to provide a available programming platform to researchers. NOX is the first and widely used controller platform providing a range of basic interface. Users can access, control, and manage the global network information, and program network applications through the interfaces provided by NOX. With the expansion of SDN, the processing capacity of centralized controller with single structure, such as NOX, is limited, while the extension is difficult and encounter a performance bottleneck. Therefore, this controller is only available for small business networks or research simulation. There are two ways to expand the single centralized controller: improving the controller's ability or implementing multiple controllers to improve the overall processing power.

The controller has the global network information for processing, so it is required to have a high processing capacity. NOX-MT is a NOX controller with the capability for processing multithreading, which improves the performance of NOX [35]. NOX-MT does not change the basic structure of NOX controller, but improves the performance by the traditional parallel technology. Thus, the users can quickly update NOX to NOX-MT without the inconsistencies caused by the platform replacement. Furthermore, Maestro is another representative parallel controller to give full play to high performance of multi-core parallel processing, which is better than NOX for processing a large-scale network traffic [25].

For the medium-sized networks, a single controller is usually competent without a significant impact on the performance. However, for the large-scale networks, it is necessary to implement multi-threaded approach to guarantee the performance. A large-scale network can be divided into a number of domains, as shown in Fig. 4.6. If a single controller processes the switch request through the centralized control, there will be a significant delay between the controller and the switches in other domains. Furthermore, it will affect the network processing performance. With the

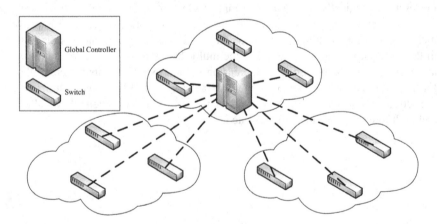

**Fig. 4.6** Single controller in SDN

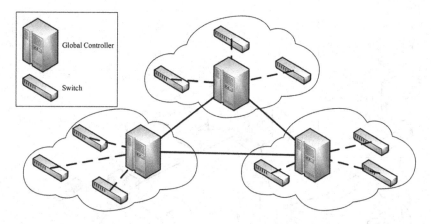

**Fig. 4.7**  Flat controller in SDN

further network expansion, the impact will become insufferable. In addition, the single node failure issue is possible in the single centralized control, which can be solved by increasing the number of the controllers, i.e., the controllers are physically distributed in the network with the logically centralized control. This allows each switch to interact with the controller for enhancing the overall performance of the network.

The distributed controllers typically include two modes for extension: the flat control (as shown in Fig. 4.7) and the hierarchical control (as shown in Fig. 4.8). For the flat control mode, all controllers are placed in the disjoint regions and, respectively, manage their own network. Each controller is identical, and communicate with the others through the east-westbound interfaces. For the hierarchical control, the vertical management functions are available between the controllers, i.e., the local controller is responsible for its own network while the global controller is responsible for the local controllers. The interactions among the controllers can be completed by the global controller.

With the flat control mode, all the controllers are placed at the same level. Although each controller is physically located in different areas, all the controls are logically as same as the global controller for managing the global network status. When the network topology is changed, all the controllers are updated simultaneously, and the switch only needs to adjust the address mapping of the controller without other complex update operation. Therefore, the flat distributed expansion affects little on the data layer. Onix is the first distributed SDN controller supporting the flat distributed controller architecture, which is managed by network information base (NIB) [19]. Each controller has a corresponding NIB, while the controllers synchronize updating is based on the consistency of NIB. HyperFlow is a distributed control plane for OpenFlow, allowing the network operators to freely deploy multiple controllers at every region of the network [34]. The controllers are physically separated but logically centralized as same as SDN centralized

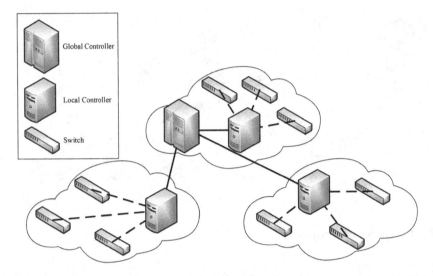

**Fig. 4.8** Hierarchical controller in SDN

control. HyperFlow communicates through registering and broadcast mechanisms, and when a controller fails, the switch controlled by the failed controller is manually reconfigured to be controlled by a new controller for ensuring the availability. In the flat control mode, although each controller manages the global network status, it only controls the local network, which results in a waste of some resources and increasing overall network load for updating the controller. In addition, in the actual deployment, the different domains may have different economic entities operators, so it is impossible to realize the equal communication between the controllers in different domains.

In the hierarchical control mode, the controllers are classified according to their usage. The local controller is relatively close to the switch, which is responsible for controlling the nodes and manage the network status in the region. For example, the controller close to the switch only supports for the regular interaction and high hit rules distribution while the global controller is responsible for the maintenance of the entire network information and complete routing and other operations based on the entire network information. The interaction of hierarchical controller includes two model: the interaction between the local controller and the global controller, and the interaction among the global controllers. For the domains belonged to the different operators, only the interaction among the global controllers should be negotiated. Kandoo realizes the hierarchical distributed structure [15]. When the switch is forwarding packets, it firstly asks the local controller nearby. If the packet is the local information, the local controller will respond quickly. If the local controller cannot process the message, it will ask the global controller and send the obtained information to the switch. It avoids the frequent interaction of the global controllers and effectively reduces the traffic load. Because it depends on the hit information processed by the local controller, the scenario involving more local applications has a higher efficiency.

**Interface**

The controller provides the users with the NBIs for the user-friendly network configuration. However, with the various network applications, such as traffic monitoring, load balancing, access control, and routing, the traditional controller platform, such as NOX, can only provide low-level configuration interface programmed by the common language, such as $C^{++}$, which is a lower level abstraction and the costs for the network deployment is not significantly decreased. Hence, an high-level abstracted configuration language is expected to unify the NBI and improve the performance of the interface for reducing the overall cost of the network configuration.

Yale University has developed a series of network configuration language to build a common NBIs for optimizing the performance. Nettle is proposed as a descriptive language using the functional reactive programming (FRP) mode [37]. FRP is an approach based on event-driven programming, which is suitable for the controller responding to the various applications. Nettle is expected to support a programmable approach for the network configuration, but the programmable network configuration requires a high performance provided by the controller. Therefore, McNettle is proposed as a multi-core Nettle language, which does not change the descriptive characteristics of FRP for improving the user experience of the development through the shared memory and multi-core processing [36]. Moreover, Procera is an optimized abstractive language using the advanced network strategy for all applications [38]. In order to improve the performance of the controller interface language, Maple is proposed [39], as shown in Fig. 4.9. Maple provides the users with the customized abstracting policy. In order to the effectively decomposing the abstracting policy into a series of rules and distributing to the corresponding switch, Maple uses a highly efficient multi-core scheduler, and even uses a tracing runtime optimizer to optimize the performance. The optimizer transfer the load to the switches as much as possible by recording the reusable policy, while the flow table is up to date by dynamic tracking abstracting strategy and the dependence between the data content and the environment for ensuring the efficiency of transforming the abstracting strategies to the available rules.

Frenetic proposed by the joint research group of Cornell University and Princeton University, is similar to Nettle and other languages that it is descriptive and functional reactive [11]. However, it includes the following three difference between Frenetic and Nettle:

1. Since Frenetic is based on NOX, so its level of abstraction is higher than Nettle.
2. Frenetic includes a query language and real-time system, which is not supported by Nettle.
3. Frenetic is packet-based, while Nettle is event-stream-based.

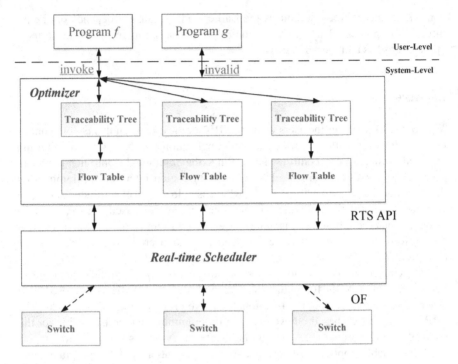

**Fig. 4.9** Maple

Furthermore, NetCore [24] enhances the ability of Frenetic, including adding wildcard matching mode, automatically generating rules and common grammatical structures. Since Frenetic and NetCore are based on the parallel modular assembly mode, which requires each module, i.e., service, such as access control is required to generate a backup packet, resulting in some resources wastes. Pyretic is based on a sequential modular assembly mode, which makes each application be processed in order, and effectively avoid the possible conflict caused by the potential applications [30]. In order to enhance the theoretical foundation for such interface languages, NetKAT is proposed to use Kleene algebra for detection, which improves NetCore that the mathematical theories are available to verify the correctness of configuration and avoid the potential problems of network configuration [3].

**The Features of Control Layer**

Consistency, availability, and fault tolerance are the most important features of control layer, but these three features cannot be satisfied at the same time. Therefore, in major cases, researchers focus on partly improving the features with a little negative influence on the other features. This subsection discusses the consistency, availability, and fault tolerance, respectively.

1. **Consistency**: Compared with other network architectures, centralized control is one of the most significant advantages of SDN. Through the centralized control, the user can obtain a global view of the network information for the unified network design and deployment, which ensures the consistency of the network configuration. However, the distributed controllers may still causes the potential inconsistencies. Due to the different design of the controllers, the requirements of the consistency are different. Therefore, the controller keeping the distributed status unified globally cannot guarantee the network performance. On the other hand, if the controller is able to respond quickly to the requests and distribute the policy, the global status consistency is not guaranteed. Hence, it is a great challenge to ensure the consistency without a significant influence on the performance for SDN.

   Similarly, the concurrency strategy may also cause the consistency issue, which can be solved by the control layer making the rules and submitting in a two-procedure way. In order to avoid the excessive involvement of the data layer, the control layer directly concurrently combines the rules and ensures no rules conflicts through the fine-grained locking. For example, Hierarchical Flow Tables (HFT) is proposed for specifying and realizing hierarchical policies in SDN. HFT policies are organized as trees, where each component of the tree can independently determine the action to take on each packet. When independent parts of the tree arrive at conflicting decisions, HFT resolves conflicts with user-defined conflict-resolution operators, which exist at each node of the tree [10].

2. **Availability**: Rules backup can enhance network availability. For example, RuleBricks is proposed for flexibly embedding high availability support in existing OpenFlow policies. RuleBricks introduces three key primitives: drop, insert, and reduce, which can express various flow assignment and backup policies. It verifies that RuleBricks maintains linear scalability with the number of replicas on the Chord ring [40].

   SDN controller is the core processing nodes to handle a large number of requests from the switch, while the heavy load may affect the availability of SDN. The distributed controller is available for load balance and improving the overall performance of SDN. In particular, for the hierarchical controller, such as Kandoo, the local controller is used to process the major requests from the switch, while the global controller can provide better services for users. However, there is also an usability issue in the distributed controller architecture. Since each controller needs to deal with different switches and the network traffic distribution is not balance, so the availability of some controllers is decreased. To address this problem, ElastiCon is proposed, which is an elastic distributed controller architecture in which the controller pool is dynamically grown or shrunk according to traffic conditions. To address the load imbalance caused due to spatial and temporal variations in the traffic conditions, ElastiCon automatically balances the load across controllers thus ensuring good performance at all times irrespective of the traffic dynamics [8].

   Furthermore, the availability of the control layer is able to be improved by reducing the number of requests from the switch. For example, DIFANE, a scalable and efficient solution, is proposed to keep all traffic in the data plane by selectively directing packets through intermediate switches that store the necessary rules. DIFANE relegates the controller to the simpler task of partitioning these rules over the switches. DIFANE can be readily implemented with commodity switch hardware, since all data-plane functions can be expressed in terms of wildcard rules that perform simple actions on matching packets [43]. DevoFlow, a modification of the OpenFlow model, is proposed to gently break the coupling between control and global visibility, in a way that maintains a useful amount of visibility without imposing unnecessary costs. DevoFlow can load-balance data center traffic as well as fine-grained solutions, without as much overhead: DevoFlow uses 10–53 times fewer flow table entries at an average switch, and uses 10–42 times fewer control messages [7].

3. **Fault Tolerance**: Similar to the traditional Internet, SDN also facing the network node or link failure. Fortunately, SDN controller can quickly restore the node failure though the entire network information, illustrating a strong ability of fault tolerance. In Fig. 4.10, it shows the network node recovery convergence process: (1) A switch's failure is detected by the other switches; (2) The switch notifies the controller with the failure; (3) The controller calculates the rules needing to restore based on the available information; (4) The updates are sent to the data plane network elements affected; (5) The data plane elements affected updates the flow table information, respectively.

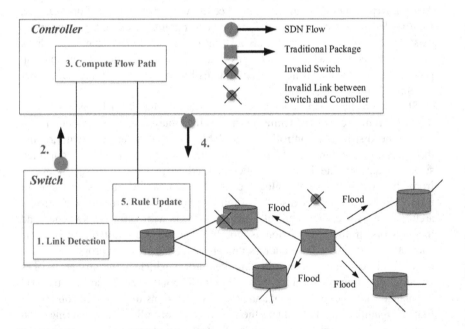

**Fig. 4.10** Convergence of failed node or link

From the link recovery process, it can be seen that the failure information in the SDN architecture is not informed the whole network by flooding, but directly sent to the control layer and the controller makes the recovery decision, which is possible to cause the routing oscillation phenomenon. If the link between the switch and the controller fails resulting in unable communication, it will be relatively difficult for the convergence. It can be recovery by flooding through the Interior Gateway Protocol (IGP) of the traditional network. Moreover, failover is also available to alleviate the link failure convergence time. Through the switch installed with the static forwarding rules to verify the topology connectivity, the fast convergence of network failures is more available.

To avoid the node failure caused by manual configuring, the control layer provides a senior network fault tolerance language FatTire. The central feature of this language is a new programming construct based on regular expressions that allows developers to specify the set of paths that packets may take through the network as well as the degree of fault tolerance required. This construct is implemented by a compiler that targets the in-network fast-failover mechanisms provided in recent versions of the OpenFlow standard, and facilitates simple reasoning about network programs even in the presence of failures [31].

### 4.3.4   SDN-Based Application

With the rapid development of SDN, SDN has been applied to various network scenarios, from small enterprise network and campus network to the data center (DC) and the Wide Area Network (WAN), expanding from the wired network to the wireless network. Regardless of the scenario, the separated control layer and data layer of SDN are implemented in most applications to obtain a global view for managing the networks.

#### Enterprise Networks and Campus Networks

The early SDN-based applications are often deployed in the enterprise network or campus network. However, in actual deployment later, due to the different demands of the enterprises or campus, it is difficult to deploy SDN in accordance with its own characteristics. To address this issue, the SDN is improved to support for a customized management of the enterprise networks and campus network. For example, Hyojoon and Nick propose an improved network management with SDN for enabling frequent changes to network conditions and state, providing support for network configuration in a high-level language, and providing better visibility and control over tasks for performing network diagnosis and troubleshooting [18]. Furthermore, the consistency of network deployment also attracts attention, because the forwarding topology looping and invalid configuration and other issues are possible through the customized SDN management. Therefore, OF.CPP is proposed to use transactional semantics at the controller to achieve consistent packet processing [27].


**Data Center and Cloud**

Due to the complicated equipment and highly concentrated, it is also a great challenge to deploy SDN in the DC. The early SDN deployments in the DC are based on NOX, and then SDN-base applications are widely deployed in the DC. In particular, the performance and energy efficiency are often considered in the deployment.

Thousands of computers in the DC require adequate bandwidth, so there are various issues to increase the utilization use of bandwidth, save resources, improve performance, etc. For example, CrossRoads, a network fabric, is proposed to provide layer agnostic and seamless live and offline VM mobility across multiple data centers, which leverages SDN and implements an OpenFlow-based prototype of CrossRoads. CrossRoads extends the idea of location independence based on pseudo addresses with a control plane overlay of OpenFlow network controllers in various data centers [21]. Hedera a scalable, dynamic flow scheduling system is proposed to adaptively schedule a multi-stage switching fabric to efficiently utilize aggregate network resources. It evaluates that Hedera delivers bisection bandwidth that is 96 % of optimal and up to 113 % better than static load-balancing methods [1]. Moreover, zUpdate is proposed to perform congestion-free network updates under asynchronous switch and traffic matrix changes in the DC [20].

Energy conservation is an issue of DC cannot be ignored. Generally, the large-scale stable and efficient Internet services provided by the DC are at the expense of energy wastage. Although it is able to save a small amount of energy by temporarily closing the ports with no traffic, the most effective way is to manage the global information through SDN and closed the idle devices in real time for saving about 50 % of the energy consumption. Furthermore, the low utilization also results in the higher energy consumption of DC. In the DC, each stream exclusively occupies the route in each time slice, which effectively improves the utilization of the route. With the full use of network information provided by SDN, each stream is fairly scheduled, so the route is fully utilized and the energy is saved.

Through the DC, the users can conveniently manage network via cloud. However, the topology of cloud-based network is variable. Fortunately, through SDN, the global management of the cloud-based network is available. For example, IBM proposes a dynamic graph query primitives for SDN-based cloudnetwork management, including a shared graph library that can support network management operations. Using the illustrative case of all pair shortest path algorithm, we demonstrate how scalable lightweight dynamic graph query mechanisms can be implemented to enable practical computation times, in presence of network dynamism [29].

**WAN**

WAN connects to many DCs, to provide efficient connection and traffic transmission. In order to provide reliable services, it should ensure that any link or routing failure could not affect the efficient operation of the network. Furthermore, the stability of the traditional WAN is at the expense of the link utilization, so that the average utilization of the WAN is only 30–40 %, while it is only 40–60 % under heavy network load [44].

In order to improve the utilization, Google proposes B4 to provide a private WAN connecting Google's data centers across the planet. B4 has a number of unique characteristics:

1. Massive bandwidth requirements deployed to a modest number of sites;
2. Elastic traffic demand that seeks to maximize average bandwidth;
3. Full control over the edge servers and network, which enables rate limiting and demand measurement at the edge.

B4's centralized traffic engineering service drives links to near 100 % utilization, while splitting application flows among multiple paths to balance capacity against application priority/demands [17].

Furthermore, SWAN is proposed by Microsoft to boost the utilization of inter-datacenter networks by centrally controlling when and how much traffic each service sends and frequently re-configuring the network's data plane to match current traffic demand. But done simplistically, these re-configurations can also cause severe, transient congestion because different switches may apply updates at different times. A novel technique is developed to leverage a small amount of scratch capacity on links to apply updates in a provably congestion-free manner, without making any assumptions about the order and timing of updates at individual switches. Further, to scale to large networks in the face of limited forwarding table capacity, SWAN greedily selects a small set of entries that can best satisfy current demand. It updates this set without disrupting traffic by leveraging a small amount of scratch capacity in forwarding tables [16].

**Wireless Networking**

Early SDN technology research has been deployed in the wireless network, which has been widely used in various aspects of a wireless network. For example, Open-Roads, an open-source platform for innovation in mobile networks, is proposed to enable researchers to innovate using their own production networks, through providing an wireless extension OpenFlow [41]. Odin, an SDN framework to introduce programmability in enterprise wireless local area networks (WLANs), is proposed to support a wide range of services and functionalities, including authentication, authorization and accounting, policy, mobility and interference management, and load balancing [33]. OpenRadio is proposed for a programmable wireless dataplane that provides modular and declarative programming interfaces across the entire wireless stack [4]. SoftRAN, a fundamental rethink of the radio access layer, is proposed. SoftRAN is a software defined centralized control plane for radio access networks that abstracts all base stations in a local geographical area as a virtual big-base station comprised of a central controller and radio elements (individual physical base stations) [14].

## 4.4  Networking as a Service

The concept of Networking as a Service (NaaS) is originated in the cloud computing, which is expected to provide on-demand enterprise network services for creating a virtual network to support the applications and provide the flexible, scalable design and deployment of network services function [45]. Compared with the conventional network providing the connection for the fixed location, server to server, and server to storage, NaaS is significantly different. In some ways, this is just to change the business model, because the suppliers still need to deploy the same level of infrastructure. But from the customer's point of view, the user can pay a monthly fee based on their demands, while NaaS enables them to use the network services without any infrastructure deployment.

The first NaaS-based application is the virtual overlay network. The network software provider Nicira, acquired by VMWare in 2012,[6] developed Nicira NVP,[7] a completely software-based virtual network architecture can be controlled by the cloud API. Virtual overlay network is a combination of tunneling and virtual switches, it create a network beyond the physical network equipments, such as switches and routers.

Because NaaS is deployed and controlled by the software, the application network is a essentially transparent appendage for the actual network equipment and services creating the connection. Virtual overlay network classifies the connection created by the actual services and infrastructures, but it does not create any connection or change the contractual relationship of the service, which is a perfect solution for dynamically controlling the network applications without any change to the site connection, construction mode, and the content of service.

### 4.4.1  Create a Virtual Network Segment

If the network services need to become more flexible and application-driven, more developments are necessary. One of the most important limitations of virtual overlay network is that it is impossible to create a actual network, though NaaS is considered

---

[6]https://en.wikipedia.org/wiki/Nicira.

[7]https://en.wikipedia.org/wiki/Network_virtualization_platform.

to be a novel flexible network service. In fact, the virtual overlay can only distinguish the connections provided by the actual network or service.

The virtual overlay network is not the only way to create virtual network segments. In Ethernet networks, the actual infrastructure can be divided into virtual networks through VLAN, Virtual Extensible LAN (VXLAN) and other standards, and the network operators provide VLAN and virtual private network (VPN) services.

The collaboration between the software/cloud and the NMS is the most significant disadvantage to create NaaS. Though the collaboration is available, it is important to ensure that the network and services contains all the key elements in the network applications. The application of virtual network often requires Dynamic Host Configuration Protocol (DHCP) server for address allocation, Domain Name System (DNS) for address resolution, and default gateway, i.e., router, to connect the application network to the user.

### 4.4.2 Integration of NaaS and WAN

NaaS still requires a integration with a static set of underlying WAN services. Therefore, NaaS model is not considered to be completely flexible. All traffic of NaaS must be in the range of the actual network capacity, including WAN link. Someone believes that NaaS does not integrate with the device, which makes it more difficult to manage. NMS tools just regards NaaS as traffic. Therefore, the full flexibility of NaaS means that it needs to integrate with WAN.

### 4.4.3 Advantage of NaaS

Compared with the general network solutions, NaaS brings some advantages. For example, the customers can get the operational advantages from a policy-based centralized traffic control, and they can also enjoy greater flexibility, resource optimization, scalability, network efficiencies, and cost savings. In addition, NaaS also provides greater analytical capabilities and disaster recovery options, which are difficult to deploy in a separate configuration.

## References

1. M. Al-Fares, S. Radhakrishnan, B. Raghavan, N. Huang, A. Vahdat, Hedera: dynamic flow scheduling for data center networks, in *NSDI*, vol. 10 (2010), pp. 19–19
2. C. Amza, G. Soundararajan, E. Cecchet, Transparent caching with strong consistency in dynamic content web sites, in *Proceedings of the 19th Annual International Conference on Supercomputing* (ACM, New York, 2005), pp. 264–273

3. C.J. Anderson, N. Foster, A. Guha, J.-B. Jeannin, D. Kozen, C. Schlesinger, D. Walker, Netkat: semantic foundations for networks. ACM SIGPLAN Not. **49**(1), 113–126 (2014)

4. M. Bansal, J. Mehlman, S. Katti, P. Levis, Openradio: a programmable wireless dataplane, in *Proceedings of the First Workshop on Hot Topics in Software Defined Networks* (ACM, New York, 2012), pp. 109–114

5. G. Barish, K. Obraczke, World wide web caching: trends and techniques. IEEE Commun. Mag. **38**(5), 178–184 (2000)

6. P. Bosshart, G. Gibb, H.-S. Kim, G. Varghese, N. McKeown, M. Izzard, F. Mujica, M. Horowitz, Forwarding metamorphosis: fast programmable match-action processing in hardware for SDN, in *ACM SIGCOMM Computer Communication Review*, vol. 43 (ACM, New York, 2013), pp. 99–110

7. A.R. Curtis, J.C. Mogul, J. Tourrilhes, P. Yalagandula, P. Sharma, S. Banerjee, Devoflow: scaling flow management for high-performance networks. ACM SIGCOMM Comput. Commun. Rev. **41**(4), 254–265 (2011)

8. A.A. Dixit, F. Hao, S. Mukherjee, T. Lakshman, R. Kompella, Elasticon: an elastic distributed sdn controller, in *Proceedings of the Tenth ACM/IEEE Symposium on Architectures for Networking and Communications Systems* (ACM, New York, 2014), pp. 17–28

9. A. Dutta, H. Schulzrinne, Marconinet: overlay mobile content distribution network. IEEE Commun. Mag. **42**(2), 64–75 (2004)

10. A.D. Ferguson, A. Guha, C. Liang, R. Fonseca, S. Krishnamurthi, Hierarchical policies for software defined networks, in *Proceedings of the First Workshop on Hot Topics in Software Defined Networks* (ACM, New York, 2012), pp. 37–42

11. N. Foster, R. Harrison, M.J. Freedman, C. Monsanto, J. Rexford, A. Story, D. Walker, Frenetic: a network programming language, in *ACM Sigplan Notices*, vol. 46 (ACM, New York, 2011), pp. 279–291

12. S. Ghorbani, M. Caesar, Walk the line: consistent network updates with bandwidth guarantees, in *Proceedings of the First Workshop on Hot Topics in Software Defined Networks* (ACM, New York, 2012), pp. 67–72

13. A. Greenberg, G. Hjalmtysson, D.A. Maltz, A. Myers, J. Rexford, G. Xie, H. Yan, J. Zhan, H. Zhang, A clean slate 4d approach to network control and management. ACM SIGCOMM Comput. Commun. Rev. **35**(5), 41–54 (2005)

14. A. Gudipati, D. Perry, L.E. Li, S. Katti, SoftRAN: software defined radio access network, in *Proceedings of the Second ACM SIGCOMM Workshop on Hot Topics in Software Defined Networking* (ACM, New York, 2013), pp. 25–30

15. S. Hassas Yeganeh, Y. Ganjali, Kandoo: a framework for efficient and scalable offloading of control applications, in *Proceedings of the First Workshop on Hot Topics in Software Defined Networks* (ACM, New York, 2012), pp. 19–24

16. C.-Y. Hong, S. Kandula, R. Mahajan, M. Zhang, V. Gill, M. Nanduri, R. Wattenhofer, Achieving high utilization with software-driven WAN, in *ACM SIGCOMM Computer Communication Review*, vol. 43 (ACM, New York, 2013), pp. 15–26

17. S. Jain, A. Kumar, S. Mandal, J. Ong, L. Poutievski, A. Singh, S. Venkata, J. Wanderer, J. Zhou, M. Zhu et al., B4: experience with a globally-deployed software defined wan. ACM SIGCOMM Comput. Commun. Rev. **43**(4), 3–14 (2013)

18. H. Kim, N. Feamster, Improving network management with software defined networking. IEEE Commun. Mag. **51**(2), 114–119 (2013)

19. T. Koponen, M. Casado, N. Gude, J. Stribling, L. Poutievski, M. Zhu, R. Ramanathan, Y. Iwata, H. Inoue, T. Hama et al., Onix: a distributed control platform for large-scale production networks, in *OSDI*, vol. 10 (2010), pp. 1–6

20. H.H. Liu, X. Wu, M. Zhang, L. Yuan, R. Wattenhofer, D. Maltz, zUpdate: updating data center networks with zero loss, in *ACM SIGCOMM Computer Communication Review*, vol. 43 (ACM, New York, 2013), pp. 411–422

21. V. Mann, A. Vishnoi, K. Kannan, S. Kalyanaraman, CrossRoads: seamless VM mobility across data centers through software defined networking, in *2012 IEEE Network Operations and Management Symposium* (IEEE, Maui, 2012), pp. 88–96
22. R. McGeer, A safe, efficient update protocol for OpenFlow networks, in *Proceedings of the First Workshop on Hot Topics in Software Defined Networks* (ACM, New York, 2012), p. 61–66
23. N. McKeown, T. Anderson, H. Balakrishnan, G. Parulkar, L. Peterson, J. Rexford, S. Shenker, J. Turner, Openflow: enabling innovation in campus networks. ACM SIGCOMM Comput. Commun. Rev. **38**(2), 69–74 (2008)
24. C. Monsanto, N. Foster, R. Harrison, D. Walker, A compiler and run-time system for network programming languages, in *ACM SIGPLAN Notices*, vol. 47 (ACM, New York, 2012), pp. 217–230
25. E. Ng, *Maestro: A System for Scalable Openflow Control* (Rice University, Houston, TX, 2010)
26. H. Pan, H. Guan, J. Liu, W. Ding, C. Lin, G. Xie, The FlowAdapter: enable flexible multi-table processing on legacy hardware, in *Proceedings of the Second ACM SIGCOMM Workshop on Hot Topics in Software Defined Networking* (ACM, New York, 2013), pp. 85–90
27. P. Perešíni, M. Kuzniar, N. Vasić, M. Canini, D. Kostiū, OF. CPP: consistent packet processing for OpenFlow, in *Proceedings of the Second ACM SIGCOMM Workshop on Hot Topics in Software Defined Networking* (ACM, New York, 2013), pp. 97–102
28. M. Rabinovich, O. Spatscheck, *Web Caching and Replication* (Addison-Wesley, Boston, 2002)
29. R. Raghavendra, J. Lobo, K.-W. Lee, Dynamic graph query primitives for sdn-based cloudnetwork management, in *Proceedings of the First Workshop on Hot Topics in Software Defined Networks* (ACM, New York, 2012), pp. 97–102
30. J. Reich, C. Monsanto, N. Foster, J. Rexford, D. Walker, Modular sdn programming with pyretic. Technical Report of USENIX (2013)
31. M. Reitblatt, M. Canini, A. Guha, N. Foster, Fattire: declarative fault tolerance for software-defined networks, in *Proceedings of the Second ACM SIGCOMM Workshop on Hot Topics in Software Defined Networking* (ACM, New York, 2013), pp. 109–114
32. L. Rizzo, L. Vicisano, Replacement policies for a proxy cache. IEEE/ACM Trans. Netw. **8**(2), 158–170 (2000)
33. L. Suresh, J. Schulz-Zander, R. Merz, A. Feldmann, T. Vazao, Towards programmable enterprise WLANS with Odin, in *Proceedings of the First Workshop on Hot Topics in Software Defined Networks* (ACM, New York, 2012), pp. 115–120
34. A. Tootoonchian, Y. Ganjali, HyperFlow: a distributed control plane for OpenFlow, in *Proceedings of the 2010 Internet Network Management Conference on Research on Enterprise Networking* (2010), pp. 3–3
35. A. Tootoonchian, S. Gorbunov, Y. Ganjali, M. Casado, R. Sherwood, On controller performance in software-defined networks, in *Presented as Part of the 2nd USENIX Workshop on Hot Topics in Management of Internet, Cloud, and Enterprise Networks and Services* (2012)
36. A. Voellmy, J. Wang, Scalable software defined network controllers, in *Proceedings of the ACM SIGCOMM 2012 Conference on Applications, Technologies, Architectures, and Protocols for Computer Communication* (ACM, New York, 2012), pp. 289–290
37. A. Voellmy, A. Agarwal, P. Hudak, Nettle: functional reactive programming for openflow networks. Technical report, DTIC Document (2010)
38. A. Voellmy, H. Kim, N. Feamster, Procera: a language for high-level reactive network control, in *Proceedings of the First Workshop on Hot Topics in Software Defined Networks* (ACM, New York, 2012), pp. 43–48
39. A. Voellmy, J. Wang, Y.R. Yang, B. Ford, P. Hudak, Maple: simplifying sdn programming using algorithmic policies. ACM SIGCOMM Comput. Commun. Rev. **43**(4), 87–98 (2013)
40. D. Williams, H. Jamjoom, Cementing high availability in OpenFlow with RuleBricks, in *Proceedings of the Second ACM SIGCOMM Workshop on Hot Topics in Software Defined Networking* (ACM, New York, 2013), pp. 139–144

41. K.-K. Yap, M. Kobayashi, R. Sherwood, T.-Y. Huang, M. Chan, N. Handigol, N. McKeown, Openroads: empowering research in mobile networks. ACM SIGCOMM Comput. Commun. Rev. **40**(1), 125–126 (2010)
42. F.Z. Yousaf, M. Liebsch, A. Maeder, S. Schmid, Mobile cdn enhancements for qoe-improved content delivery in mobile operator networks. IEEE Netw. **27**(2), 14–21 (2013)
43. M. Yu, J. Rexford, M.J. Freedman, J. Wang, Scalable flow-based networking with difane. ACM SIGCOMM Comput. Commun. Rev. **40**(4), 351–362 (2010)
44. C.-K. Zhang, Y. Cui, H.-Y. Tang, J.-P. Wu, State-of-the-art survey on software-defined networking (sdn). J. Softw. **1**(01), 62–81 (2015)
45. Z. Zirui, W. Jingyu, X. Tong, Tenant-oriented customized vm networking technology. Telecommun. Sci. **31**(10), 2015267 (2015)

# Chapter 5
# Green Wireless Networks

**Abstract** In order to improve the resource and energy efficiency of the mobile cellular networks, the architecture of green wireless networks are proposed. Based on the advanced technologies of cognitive engine with learning and decision capabilities as well as the interaction with SDN controller, a complete intelligent SDN system is available. New networking convergence architecture for future heterogeneous cellular networks based on the cognitive SDN can be expected. Wireless access network and core network will be unified under this novel architecture. In wireless access area, this chapter investigates the technology of control and data decoupling, uplink and downlink decoupling, dynamic resource adaptation with the control, and coordination of intelligent SDN controller, increasing both spectrum efficiency and energy efficiency.

## 5.1 Background

In recent years, it reports that ICT consumes a considerable part of the energy consumption all over the world. With the explosion of the business, its energy consumption is increasing. At present, ICT already consumes 3–4 % of global electricity, and this proportion is still growing at double the rate per decade [8]. In 2008, the world's largest information Technology Forum "CeBit 2 Forum" has predicted that the total energy consumption of ICT will be equivalent to the energy consumption of the entire aviation industry.

Currently, there are more than 4 million base stations to provide services to mobile users, the annual average power consumption of each base station is 25 MWh [7], i.e., the power consumption is more than 1000 billion kWh per year, which needs to consume about 36 million ton coal per year and produces over 940 million tons $CO_2$ and 30 million tons $SO_2$. Meanwhile, in order to meet the exponential growth of the user demand for data, the cellular network, local area network wireless network, body area network, and other wireless networks are growing at a similar pace, and this trend will continue for the foreseeable future. Wireless data services and the rapid increasing number of the base stations cause an incredible growth in energy consumption, leading to serious environmental pollution. According to statistics, in 2007 ICT has generated 2–3 % of the global

© The Author(s) 2016
Y. Zhang, M. Chen, *Cloud Based 5G Wireless Networks*, SpringerBriefs
in Computer Science, DOI 10.1007/978-3-319-47343-7_5

$CO_2$ emissions, while Smart 2020 reported that in 2020 the $CO_2$ emissions by the mobile communications will be three times as 2007 [27].

In order to protect the common homeland of mankind, energy conservation has become the world's most pressing common duty for the sustained economic development. To address this new challenge, Europe and other countries made it clear that the new energy economy is the future direction of the development, and propose a series of green communications as major national scientific research plan.

For the international standardization, ETSI has proposed the green agenda, including the following energy efficiency standards for the telecommunication equipment and related infrastructure [4]:

1. DTR/EE-00004 (TR 102 532): alternative energy for the telecommunications equipment.
2. DTR/EE-00002 (TR 102 530): the energy efficient architecture for the telecommunications equipment and related infrastructure.
3. DTR/EE-00003 (TR 102 531): the improved power supply and energy consumption decisions.
4. DTS/EE-00005 (TS 102 533): the energy consumption of the telecommunications equipment.

Moreover, ITU starts the research on "ICTs and Climate Change" focusing on the definition, principle, and method of the ICTs' influence on the climate change, which is expected to decrease the energy consumption of the communication networks through the following approaches [10]:

1. Reducing the number of switch and router.
2. Developing more devices adapting to climate change, e.g., utilizing natural cooling rather than air conditioning.
3. Saving energy through the existing protocols, such as the multi-energy mode of VDSL2 [6].

On the other hand, with the trends of the large-scale wireless coverage, various access terminals, the different capacity demands of the various applications, the deployment of radio access network becomes heterogeneous, fusional, and diversity [11]. In order to meet the required coverage and continuous evolution of 5G wireless networks, it is essential to design a novel network and system architecture from a higher level and a wilder view, such as the implementation of intelligent and dynamic control mechanisms, the fusion of the wireless access network and the core network, the optimization of the radio access mode, the dynamic resource allocation, etc., for optimizing the energy efficiency [2]. Therefore, with the advantages of SDN, such as the separation of control and data layers, centralized management and scheduling, and open APIs, various SDN-based approaches are proposed for the green wireless networks.

For example, in [12], an SDN-based solution is proposed to energy-aware flow scheduling, i.e., scheduling flows in the time dimension and using exclusive routing (EXR) for each flow, i.e., a flow always exclusively utilizes the links of its routing path. The key insight is that exclusive occupation of link resources usually results in higher link utilization in high-radix data center networks, since each flow does not need to compete for the link bandwidths with others. When scheduling the flows, EXR leaves flexibility to operators to define the priorities of flows, e.g., based on flow size, flow deadline, etc. In [26], an SDN-based approach is proposed to trade the increased delay in the access section, due to the utilization of energy efficient schemes, with a reduced delay in the metro section.

Although the SDN-based green wireless network develops rapidly, the difference between the wireless network and the core network brings great challenges for centralized management, decision, and control, such as the complex scenarios, the various demands, and the increasing services. Therefore, the following evolution of the existing SDN architecture is imperative for the sustainable and energy efficient 5G wireless networks:

1. Cognitive engine should be included into the architecture of SDN for complicate control and management [18], because it is necessary to efficiently recognize the wireless environment, network status, and users. Based on the cognitive process, network status, and user behavior business demands, it is available to realize resource and energy efficiency driven business awareness and resource optimization.
2. With the full utilization of the globalized and centralized management provided by SDN, it is essential to enable the collaboration between the base stations covered by the heterogeneous networks, and innovate the access mechanism to adequately improve the resource and energy efficiency.
3. The network virtualization and fragmentation are expected, while it is more important to provide more intelligent collaborative management and control through the interaction between the NES. Moreover, the access network and core network should be seamlessly integrated through the unified management supported by SDN.

## 5.2  Cognitive SDN for Green Wireless Networks

### 5.2.1  Cognitive SDN Architecture and Technology

The current SDN architecture realizes the separation of the control and forwarding. The forwarding devices complete the data transmission according to the rules distributed by the controller, while the complicated policy control, forwarding path calculation, etc., are completed by the controller. In order to support multi-applications, SDN enables the applications to directly call the controller. The standard three-tier architecture of SDN as shown in the left part of Fig. 5.1.

With the increasing volume of the heterogeneous networks, there are more status, information, and policies to be processed, but the SDN controller becomes the bottleneck of the network system. Especially for the fixed software-based functional devices, the capacity is limited and it is not suitable to regularly upgrade. Therefore,

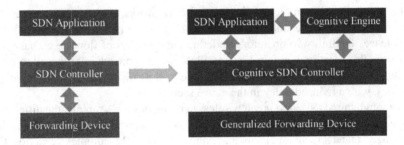

**Fig. 5.1** The evolution from SDN to cognitive SDN

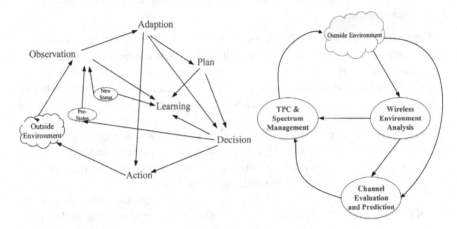

**Fig. 5.2** Cognitive radio communication system

as shown in the right side of Fig. 5.1, the cognitive engine is proposed to included into SDN for helping the controller to process information, analyze status, and provide decisions. In particular, the cognitive engine is not a conventional network device, and it is convenient to update.

As shown in Fig. 5.2, the founder of cognitive radio Dr. Joseph Mitola presented that, similar to the human brain, the cognitive radio based on the cognitive cycle is able to interact with the wireless networks. Be adaptive to the wireless environment and even the complex interference environment, the cognitive radio can autonomously change the operating characteristics and parameters and establish the corresponding cognitive cycle. Thus, all the environmental parameters sensed by the wireless network or device are available to be included into the cognitive cycle, and then the cognitive radio can provide intelligent learning, reasoning, decision making, capabilities reconfiguration, and optimization [9, 14–17].

For the cognitive SDN controller, if the cognitive engine is expected to improve the resources and energy efficiency, the energy consumption of the communication system will be reduced through the cognitive processes to create the smart green wireless communications with considerable energy efficiency and resource efficiency. Specifically, the cognitive engine in SDN can discovery more low power communication resources, reduce unnecessary idle power consumption, refine the redundant energy consumption through the cognitive process of the environment, such as network type, traffic load, channel quality, and other dynamic environmental parameters, in order to enable the green wireless communications.

The cognitive radio can lower the energy consumption as well as guarantee considerable QoS under different channel conditions. Regardless the advanced hardware, the green wireless communications need the supports by dynamic hibernation, power control, relay, dynamic spectrum management, etc., which is based on adequate environmental information. Thus, the integration of the cognitive radio and SDN will become the crucial approach for 5G wireless networks to adapt to the complex environments, dynamically effective utilize the various wireless resources, improve the resource and energy efficiency, which will provide the theoretical foundation and technical approach for the green wireless communications.

Usually, the cognitive engine is deployed on the cloud platform with big data storage, sufficient computing, and it can provide different branches of cognitive engine according to the business demands, such as user behavior analysis, traffic-oriented project, and information security, as illustrated in Fig. 5.3.

## 5.2.2 Green Wireless Network Architecture Based on Cognitive SDN

The cognitive SDN can be implemented into the heterogeneous cellular network for a comprehensive fusion, including the RAN and backbone, in order to optimize the overall network resource and improve the efficiency. The green wireless network architecture based on cognitive SDN consists of three layers, i.e., radio access layer, core transmission layer, and intelligent control layer. The radio access layer connects with the core transmission layer via the base station, while the core transmission layer interacts with the intelligent control layer through the southbound interfaces

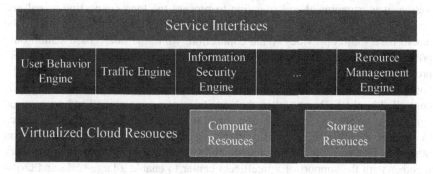

**Fig. 5.3** Cognitive engine of cognitive SDN

of SDN controller. In particular, the controller in the intelligent control layer is able to directly connect to the base station in the radio access layer for a rapid respond.

1. **Radio access layer**: The base station, with the local decision making and content storage abilities, can be considered a edge node of the radio access layer as well as the core transmission layer.
2. **Core transmission layer**: The data packet is no longer transmitted through the tunnel protocols. Based on to the flow table distributed by the cognitive SDN controller, an appropriate routing is generated according to the user's identification and business demands for providing intelligent and personalized transmission services.
3. **Intelligent control layer**: With the powerful supports by the cognitive engine, the controller is enabled to complete the computation-intensive tasks, such as context awareness, process optimization, and decision making, for evaluating the energy consumption of each base station or router, detecting the interference wireless environment, predicting the popular user content, and optimizing the network resources. In order to complete such a complex, cross-domain decision, the SDN controller collects a lot information about the user location, energy requirements, flow requirements, link bandwidth, etc., from the core transmission layer and radio access layer. Then, the dynamically loadable intelligence engine branch processes this information and provides intelligent analysis and decision making for the controller.

## 5.3  SDN-Based Energy Efficiency Optimization for RAN

### 5.3.1  Separation Between Control and Data

In users access policies of the conventional cellular network, the control signaling and data signaling related to a single user is coupled on the same base station, which causes great losses to the mobile cellular network energy efficiency [29]. Meanwhile, for the future dense deployment of small base stations or micro base station, it is economical and scientific to process the long connection signal and deal with the frequent switching. With the separation of control and data planes, the macro base station provides signal connection, while the micro base station only processes the related datas, which extensively improves the network efficiency and provide accurate and effective management. In the framework of cognitive SDN, the control base station and the data base station can be considered as a switching node controlled by the SDN controller, the cognitive engine can unified and coordinated control for signal, data transmission, base station resources, user behavior, etc., to support dynamic resource allocation, fast switching, base stations sleep, and other new features. Considering the scalability, multiple SDN controllers can work together with the support of a intelligent center to enable a large-scale and cross-domain co-optimization strategy.

### 5.3.2   Separation Between Uplink and Downlink

With the access policies of conventional cellular network, user's uplink and downlink are connected to the same base station. Generally, the appropriate base station is selected on the received signal strength indicator (RSSI) of user's downlink [23]. In the case of heterogeneous laminated cover, the same base station is often unable to guarantee optimal downlink bandwidth and optimal energy efficiency coverage. Therefore, it can be optimized through the separation between uplink and downlink which, respectively, access the different base station optimization. Specifically, the downlink often connects to the base dation with high RSSI, while the uplink directly connects the nearest base station [24]. Though the separation strategy effective optimizes the performance of uplink and downlink, and improve the energy efficient, it must be supported by the centralized control devices. In the framework of cognitive SDN, the users and base stations have to report and RSSI, channel conditions, interference distribution, user location, etc., and then the cognitive engine is able to provide the assistance for selecting the optimal access strategy and controlling the base station to provide access to the users.

### 5.3.3   Elastic Wireless Resources Matching

Faced with increasingly complex, intensive, and heterogeneous cellular network deployment, the radio interference between the base stations has become more serious, which significantly restricted the cellular network capacity and resource efficiency. Cognitive SDN-based framework is able to dynamically sense the radio resource usage, interference, energy demand and load demand from the base station, and elastically match radio resources to maximize the network throughput and performance through the intelligent decision-making. Although the user's location and the energy demand are ever-changing in the cellular network, they are still predictable [22, 25]. If it is able to accurately predict the user behavior in time and space, the limited radio resources will provide much more excellent services while the energy consumption will be significantly reduced. According to the historical data, the prediction algorithms of cognitive SDN can provide considerable prediction for the user behavior, movement, traffic, etc., and match the appropriate radio resources in advance [13, 19, 30]. In the cognitive engine, most radio resource usage can be sensed and analyzed through the installment of scalable machine learning algorithms, such as reinforcement learning [5], transfer learning [3], and deep learning [1]. With the accurate prediction of user traffic demand, it is possible to comprehensively co-optimize the network resources.

## 5.4   SDN-Based Green Wireless Networks Fusion

In the current cellular networks, the wireless and IP domains are completely separated, while the IP domain and Internet domain are completely separated. Therefore,
the wireless domain connects with the IP domain through tunnels, while the IP
domain interacts with the Internet domain via gateway, which results in inefficiency,
resources wastes, inflexible and un-unified control, and other fundamental issues.
For the future mobile cellular network, the applications need to carry huge, real-
time, and massive connections, while the tunnel and the gateway are incompetent.
Therefore, the IP domain and Internet domain have to be extended to the edge of the
cellular network. The SDN-based green wireless networks fusion mainly focuses on
the content delivery fusion based on the cognitive SDN.

Content delivery technology is one of the most important researches for the Internet, which builds an overlay network beyond the traditional Internet infrastructure,
in order to improve the user experience through CDN, cache, etc.

With the combination of CDN and cognitive SDN, it is effective to select appropriate edge server for the whole network resource optimization and dynamically
switch the edge server during the services, which greatly improves the current CDN
architecture [28]. With the fusion of content delivery and cellular network, the
essential advantage of wireless networks, i.e., broadcast channel, can be fully used
to integrate the multicast and caching mechanisms for maximizing the resources-
efficiency, energy-efficiency, and user experience [20]. In addition, considering
the user's mobility in the cellular network, a caching paradigm is proposed to
reduce the energy costs for serving the massive mobile data demand in 5G wireless
networks. In contrast to the traditional caching schemes that simply bring popular
content close to users, caching strategy is carefully designed so as to additionally
exploit multicast. The fusion of CDN and SDN is of high importance nowadays,
since multicast attracts attention as a technique for efficient content delivery in the
evolving cellular networks [21].

## References

1. M.A. Alsheikh, D. Niyato, S. Lin, H.-P. Tan, Z. Han, Mobile big data analytics using deep
   learning and apache spark. IEEE Netw. 30(3), 22–29 (2016)
2. J.G. Andrews, S. Buzzi, W. Choi, S.V. Hanly, A. Lozano, A.C. Soong, J.C. Zhang, What will
   5g be? IEEE J. Sel. Areas Commun. 32(6), 1065–1082 (2014)
3. E. Baştuğ, M. Bennis, M. Debbah, A transfer learning approach for cache-enabled wireless
   networks, in 2015 13th International Symposium on Modeling and Optimization in Mobile, Ad
   Hoc, and Wireless Networks (WiOpt) (IEEE, Mumbai, 2015), pp. 161–166
4. B. Dugerdil, Etsi green agenda. http://www.itu.int/dms_pub/itu-t/oth/06F
5. S. El-Tantawy, B. Abdulhai, H. Abdelgawad, Multiagent reinforcement learning for integrated
   network of adaptive traffic signal controllers (marlin-atsc): methodology and large-scale
   application on downtown Toronto. IEEE Trans. Intell. Transp. Syst. 14(3), 1140–1150 (2013)

6. P.-E. Eriksson, B. Odenhammar, Vdsl2: next important broadband technology. Ericsson Rev. **1**, 36–47 (2006)
7. G. Fettweis, E. Zimmermann, ICT energy consumption-trends and challenges, in *Proceedings of the 11th International Symposium on Wireless Personal Multimedia Communications*, vol. 2 (Lapland, Rovaniemi, 2008), p. 6
8. Z. Hasan, H. Boostanimehr, V.K. Bhargava, Green cellular networks: a survey, some research issues and challenges. IEEE Commun. Surv. Tutorials **13**(4), 524–540 (2011)
9. S. Haykin, Cognitive radio: brain-empowered wireless communications. IEEE J. Sel. Areas Commun. **23**(2), 201–220 (2005)
10. T. Kelly, S. Head, Icts and climate change. ITU-T Technology, Technical Report (2007)
11. B.-J.J. Kim, P.S. Henry, Directions for future cellular mobile network architecture. First Monday **17**(12) (2012)
12. D. Li, Y. Shang, C. Chen, Software defined green data center network with exclusive routing, in *IEEE INFOCOM 2014-IEEE Conference on Computer Communications* (IEEE, Toronto, 2014), pp. 1743–1751
13. R. Li, Z. Zhao, X. Zhou, J. Palicot, H. Zhang, The prediction analysis of cellular radio access network traffic: from entropy theory to networking practice. IEEE Commun. Mag. **52**(6), 234–240 (2014)
14. J. Mitola, Cognitive radio for flexible mobile multimedia communications, in *1999 IEEE International Workshop on Mobile Multimedia Communications, 1999 (MoMuC'99)* (IEEE, San Diego, 1999), pp. 3–10
15. J. Mitola III, Cognitive radio architecture, in *Cooperation in Wireless Networks: Principles and Applications* (Springer, Berlin, 2006), pp. 243–311
16. J. Mitola, Cognitive radio architecture evolution. Proc. IEEE **97**(4), 626–641 (2009)
17. J. Mitola, G.Q. Maguire, Cognitive radio: making software radios more personal. IEEE Pers. Commun. **6**(4), 13–18 (1999)
18. T.R. Newman, B.A. Barker, A.M. Wyglinski, A. Agah, J.B. Evans, G.J. Minden, Cognitive engine implementation for wireless multicarrier transceivers. Wirel. Commun. Mob. Comput. **7**(9), 1129–1142 (2007)
19. S.J. Pan, Q. Yang, A survey on transfer learning. IEEE Trans. Knowl. Data Eng. **22**(10), 1345–1359 (2010)
20. K. Poularakis, L. Tassiulas, Cooperation and information replication in wireless networks. Philos. Trans. R. Soc. A **374**(2062), 20150123 (2016)
21. K. Poularakis, G. Iosifidis, V. Sourlas, L. Tassiulas, Exploiting caching and multicast for 5g wireless networks. IEEE Trans. Wirel. Commun. **15**(4), 2995–3007 (2016)
22. J.B. Rao, A.O. Fapojuwo, A survey of energy efficient resource management techniques for multicell cellular networks. IEEE Commun. Surv. Tutorials **16**(1), 154–180 (2014)
23. S. Singh, X. Zhang, J.G. Andrews, Joint rate and sinr coverage analysis for decoupled uplink-downlink biased cell associations in hetnets. IEEE Trans. Wirel. Commun. **14**(10), 5360–5373 (2015)
24. K. Smiljkovikj, P. Popovski, L. Gavrilovska, Analysis of the decoupled access for downlink and uplink in wireless heterogeneous networks. IEEE Wireless Commun. Lett. **4**(2), 173–176 (2015)
25. J. Sun, W. Xu, B. Feng, A global search strategy of quantum-behaved particle swarm optimization, in *2004 IEEE Conference on Cybernetics and Intelligent Systems*, vol. 1 (IEEE, Singapore, 2004), pp. 111–116
26. L. Valcarenghi, K. Kondepu, A. Sgambelluri, F. Cugini, P. Castoldi, G.R. de los Santos, R.A. Morenilla, D.L. Lopez, Experimenting the integration of green optical access and metro networks based on SDN, in *2015 17th International Conference on Transparent Optical Networks (ICTON)* (IEEE, Budapest, 2015), pp. 1–4
27. M. Webb et al., Smart 2020: enabling the low carbon economy in the information age. Climate Group Lond. **1**(1), 1–1 (2008)
28. M. Wichtlhuber, R. Reinecke, D. Hausheer, An sdn-based cdn/isp collaboration architecture for managing high-volume flows. IEEE Trans. Netw. Serv. Manag. **12**(1), 48–60 (2015)

29. X. Xu, G. He, S. Zhang, Y. Chen, S. Xu, On functionality separation for green mobile networks: concept study over lte. IEEE Commun. Mag. **51**(5), 82–90 (2013)
30. X. Zhou, Z. Zhao, R. Li, Y. Zhou, H. Zhang, The predictability of cellular networks traffic, in *2012 International Symposium on Communications and Information Technologies (ISCIT)* (IEEE, Gold Coast, 2012), pp. 973–978

# Chapter 6
# 5G-Related Projects

**Abstract** This chapter introduces five representative projects on 5G, including Mobile and wireless communications Enablers for the Twenty-twenty Information Society (METIS), Multi-hop Cellular Networks (MCN), Network Function as-a-service Over Virtualized Infrastructures (T-NOVA), iJOIN and Nuage Virtualized Services Platform (NUAGE).

## 6.1 METIS

METIS, Mobile and wireless communications Enablers for the Twenty-twenty Information Society,[1] started in November 2012, which is expected to take 3 years to propose a novel 5G concepts and develop various key technologies for providing high efficiency, including energy-efficiency, spectrum-efficiency, and cost-efficiency, to the rapid growth of wireless data traffic [7].

Through METIS, it images that everyone can access information, share data, and connect to everything anytime and anywhere. Such an all connected world without information boundaries will greatly promote the development and growth of social economy. In contrast to today's wireless communication systems, 5G should significantly improve the energy consumption, cost, and resource-efficiency, which can stably improve the system capacity with the acceptable cost and energy consumption. In addition, 5G also needs versatility to support various performance requirements, such as availability, mobility, QoS, etc., and application scenarios, especially for the massive communication devices. Finally, 5G system is expected to provide considerable scalability, enabling the system to meet various demands, while keeping the efficiency of the cost, energy, and resource.

Comparing with LTE release 8,[2] the 5G system concept constructed by METIS supports 1000 times wireless data traffic, 10–100 times the number of access devices, 10–100 times user data rate, providing 10 times the battery life for the

---

[1]https://www.metis2020.com/.

[2]http://www.3gpp.org/technologies/keywords-acronyms/98-lte.

© The Author(s) 2016
Y. Zhang, M. Chen, *Cloud Based 5G Wireless Networks*, SpringerBriefs
in Computer Science, DOI 10.1007/978-3-319-47343-7_6

communication between the massive low power consumption devices and one fifth peer-to-peer delay. Specifically, in order to build 5G system concept, METIS focuses on the following four key technologies:

1. **Radio Link**: Addressing to the future requirements of various new wireless applications, METIS designs a novel wireless air interface. In particular, one of the most challenging is to support a wide range of various scenarios, from the extremely low data rate and low power consumption sensors to the ultra high-speed multimedia services. Thus, the novel technologies are necessary to be developed, such as transmit waveform, coding, modulation, and transceiver structure, in order to improve the spectral efficiency, energy efficiency, anti-interference ability, and robustness of the wireless link physical layer. In addition, the multiple access, media control, radio resource, etc., have to be redesigned to greatly enhance the efficiency of the system.

2. **Multi-node and Multi-antenna Transmission**: Multi-node and multi-antenna transmission technology are expected to improve the performance and capacity of the communication systems, which includes various research, such as the development of the performance limitation, architecture influence, algorithms and key technologies, and ultra dense networking (UDN), massive communication machines, and other new scenarios. Firstly, METIS focuses on the beamforming, space division multiple access, spatial multiplexing based massive multi-antenna applications, in order to provide high data rates and spectral efficiency, or improve link reliability, coverage, and energy efficiency. Then, the advanced multi-node coordination technology is developed to significantly improve spectral efficiency and user throughput, and improve the user wireless link quality in the harsh environment. These technologies implement the novel air interface and coordination, and integrate the advanced multi-node coordination to the actual system. Finally, the multi-hop communication and wireless network coding are deployed one or more relay nodes between the source and destination, in order to provide an efficient backhaul, improve the coverage and reliability, or transfer the information processing and energy consumption from the massive devices to the network.

3. **Multi-radio Access and Multi-layer Network**: The multi-radio access technology and multi-layer network are the foundation for the future wireless network to effectively deploy, operate, and optimize network, especially for the heterogeneous multi-layer and multi-radio access network. The multi-radio access and multi-layer network focuses on network coexistence, collaboration, and interference management, because the interference dependencies between communicating entities in a UDN are especially complex, and the device-to-device (D2D) communications greatly increase the degree of freedom for interference management. Furthermore, the management of demand, traffic, and mobility attracts the attention from MTEIS, which are very important and includes the user's location, and environmental information. The radio access technology and network layer choice are able to be extensively optimized by

fully utilizing these information. In particular, in order to minimize the signaling overhead, METIS proposes mobility management, especially the novel concept of UDN mobility management. Finally, METIS devotes to realize the network functionality, including the definition of new management interface, providing the automatic integration and management for various network nodes, and the efficient integration of nomadic cells in heterogeneous networks.

4. **Spectrum**: METIS will propose a novel concept of spectrum sharing to ensure that there is adequate spectrum for the wireless communication system after 2020. These concepts will greatly increase the number of available spectrum, and significantly improve the efficiency of spectrum. METIS will analyze the spectrum from 300 MHz to 275 GHz spectrum to identify the new spectrum resources and to understand their characteristics. Meanwhile, METIS will analyze the scenarios of the future wireless communications to understand the spectrum requirements beyond 2020. In addition, METIS will develop novel flexible spectrum sharing and management technology, to enable UDN operation at high frequencies and automatic direct D2D communication supporting for high mobility.

METIS project proposes a number of horizontal issues for the construction of 5G system concept. Each horizontal issues integrates a series of new techniques, and to provide effective solutions for one or more scenarios. So far, METIS has identified the following six horizontal issues [8]:

- **Direct D2D Communication**: Direct D2D represents two or more wireless terminals communicate directly, i.e., the user plane data does not go through the network infrastructure. Comparing to the conventional peer-to-peer communications, such as Bluetooth, the significant feature of D2D is that the links between devices are still managed by the network, including radio resource management and interference management. Therefore, D2D is able to increase network coverage, availability and reliability, distribute Backhaul, and reduce cost. In addition, D2D can improve the spectrum utilization, and enhance network capacity per area. With D2D deployment, the unnecessary network wireless link will be removed and the network resources allocation will be optimized according to the actual situation. Since the shortened distance between the transmitter and receiver, D2D can improve QoS in densely populated networks. In addition, through the development of novel resource allocation and interference management techniques, D2D will be integrated into the multidimensional network, including multi-access network and multi-layer network, which can be integrated into the entire wireless communication system.
- **Massive Machine Communication**: Massive machine communication (MMC) will be a very important component of the future wireless communication systems, which is expected to provide links for tens of billions networked device so that the network can be flexibly scaled up or down. Especially, the greatest challenge is that machine-centric communication has a wide range of unique characteristics and requirements on data rate, delay, cost, availability, and

reliability, which are quite different from the current human-centered communications. METIS will develop novel technologies based on these characteristics and needs, to support MMC and the future all connected world.

- **Moving Network**: Moving network (MN) consists of one or more nodes, while each node can be networked cars, buses, or other moving devices. In such a network, each node can communicate with surrounding nodes including fixed or mobile nodes on or off the mobile network that a large number of mobile devices are enabled to connect with each other. The development of MN is almost related to all the technical elements mentioned aboce, especially the backhaul, mobility and interference management, as well as spectrum and network sharing models and techniques.

- **UDN**: UDN will be the crucial solution for addressing to the high traffic demand beyond 2020, which is expected to increase capacity, improve the energy efficiency of wireless links and spectrum utilization. Actually, densification of infrastructure has been deployed in the existing wireless cellular network that the minimum distance between the base station is about 200 m. Within METIS, it plans to further increase the network density by several times. However, UDN will face many new technical challenges, including mobile, backhaul (especially self-backhaul), etc. Within METIS, advanced interference and mobility management will be developed at the physical and network layers to support UDN. In addition, METIS will evaluate the performance of UDN in terms of cost, energy-efficiency, spectrum utilization, etc.

- **Ultra Reliable Communication**: Ultra reliable communication (URC) is expected to increase the network availability. METIS will provide a scalable, cost-effective solutions to support high availability and reliability requirements of the service, such as telemetry services and automation control services. Currently, some purpose-built network such as public security networks have been built up, can ensure high reliability and high security. The new concepts and programs of METIS will support the evolution and integration of these networks and enable them to benefit from the economies of scale within the entire wireless communications markets.

- **Architecture**: METIS focuses on the network architecture and key techniques of the future wireless mobile communication system, and designs the entire wireless system features related to the function, topology, and interface. In other words, a novel 5G network architecture will be created, which integrates all the technical elements mentioned above.

## 6.2  Multi-hop Cellular Networks

Multi-hop Cellular Networks (MCN) are proposed as the wireless communications architecture, which combines the advantages of fixed infrastructure, i.e., base stations, and the flexibility of the ad hoc wireless network of [6]. The MCN reduces the number of base stations or limits the routing sensitivity of the ad hoc wireless network, while the throughput and performance are improved. In contrast to Single-hop Cellular Networks (SCN), many research has verified that the throughput of MCN is higher than SCN, while MCN effectively increase the transmission range [4, 5].

In the SCN, the mobile station can always connect to the base station in one hop, which is not usual in MCN. Specifically, the difference between MCN and SCN includes the following aspects:

- In the SCN, the base station and mobile station can always confect to each other in one hop within the same cell. When sending a packet, the mobile station always sends it to the same base station within the cell. If the source and destination nodes are in the same cell, the base station will directly send the packet to the destination nodes. Otherwise, the base station will forward the packet to the base station in the cell containing the destination node, and then this base station will forward this packet to the destination node in one hop.
- The MCN architecture is similar to the SCN, but it cannot guarantee the single-hop between the base station and mobile station. In the MCN, the transmission range of the base station and mobile station is smaller than SCN, so the concept of "cell" from SCN is changed to "sub-cell" in MCN. Like the ad-hoc wireless network, the key features of MCN is that if the connections between the mobile stations are available, the mobile station can directly communicate with others, which may cause multi-hop routing and is not supported in the SCN. If the source and destination nodes are in the same cell, other mobile stations can be used for the repeater, which may also cause multi-hop routing in the cell. If the source and destination are not in the same cell, the packets will be sent to the base station in source cell, and then forwarded to the destination base station. Finally, it will be sent to the destination node.

## 6.3  T-NOVA

T-NOVA (Network function as-a-service over virtualized infrastructures) is collaboratively developed by a number of companies and institutions, which is expected to an unified solution for NFVs deployment and management beyond the composite infrastructures.[3]

Specifically, T-NOVA aims to design and implement an integrated management structure, including automated trading systems for billing platform to manage, monitor, and optimize NFV architecture. T-NOVA transforms and enhances the current cloud computing management structure to elastic supply and IT infrastructure resources redistribution, and apply it to manage network functions. It also extends the concept of SDN, focusing on the Openflow technology, for efficient management of network resources, including network slicing, traffic redirection, and Qos.

---

[3]http://www.t-nova.eu/.

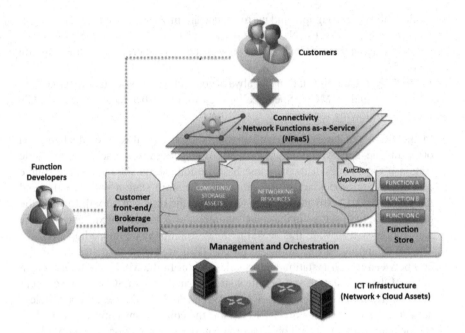

**Fig. 6.1** T-NOVA architecture (*Source*: t-nova.eu)

Except supporting the network operation/service providers to efficiently handle and manage the network, T-NOVA also proposes some innovative concepts.
Especially, it considers to provide operating users with network functions as a value-
added service, i.e., Network Functions as-a-Service (NFaaS).

As shown in Fig. 6.1, T-NOVA provides the provider with a composite service
from the NFaaS platform, consisting of the following the components:

1. Connectivity.
2. Network Functions. According to the demands, these functions includes flow
   processing, control mechanisms, and traffic offloading.

To facilitate the users T-NOVA creates an innovative "network functions store",
which is similar to the "Application Store" in some operating systems. In the
store, the network function developed the third-party developers by is published
as independent entity, including the storage and necessary metadata. Store enables
the users to select the virtual device meeting their demands to configure/modify and
plug into the existing connectivity.

To promote competition and support different configurations, a novel broker
platform, i.e., T-NOVA, is expected to allow the users to select the appropriate
services and support multiple third-party developers. After receiving the user's
request, the broker platform must check the available network and IT resources,
storage, function, etc.

Through the function store and broker platform, T-NOVA aims to promote NFV, deployment business solution, and attract more attentions from industry and academic. Especially, T-NOVA focuses the innovative network function or software, which can be included in the function store and rapidly pushed to the market, avoiding the risks causing by hardware integration and prototype delay.

## 6.4  iJOIN

iJOIN is an FP7 STREP project co-funded by the European Commission under the ICT theme (Call 8) of Directorate General for Communications Networks, Content & Technology (DG CONNECT).[4] iJOIN introduces the novel concept of RAN-as-a-Service (RANaaS), which is a centralized cloud-based open architecture for IP internet. iJOIN aims to design and optimize the backhaul connection, operation and management algorithms, and architecture elements, the integration of small cell, heterogeneous backhaul and centralized processing.

iJOIN will optimize the RAN throughput, and effectively provide services according to the dynamic demands on the cost, energy, complexity, and latency. Furthermore, with the rapid development of candidate technologies across physical, MAC, and the network layers, iJOIN will investigate the requirements, constraints and influence of the existing mobile networks, especially 3GPP LTE-A.[5]

The introduced concept of RANaaS is available to abstract new users for the RAN/backhaul market, which is similar to the cloud infrastructure or platform provider. RANaaS also provides the technical foundation for a shorter and more efficient product development cycle. Finally, iJOIN technology will significantly reduce the costs for the operators, because the computational complexity of the RAN is partly moved to the cloud infrastructure.

Specifically, iJOIN is expected to: (1) significantly improve the system throughput without increasing the spectrum resources, (2) improve the transmission efficiency by densely deploying RANaaS, (3) reduce expenses on small network deployment and operation, and (4) improve the resource utilization ratio.

## 6.5  NUAGE

NUAGE (Nuage Virtualized Services Platform) platform is a software-defined network dominated by Nokia.[6] It can virtualize any DC's network architecture and automatically connect to the computer resources at creation time. Service

---

[4]http://www.ict-ijoin.eu/.

[5]http://www.3gpp.org/technologies/keywords-acronyms/97-lte-advanced.

[6]http://www.nuagenetworks.net/.

virtualization platform based on programmable business logic and policy engine provides an open and rapid responsive product and improves the scalability for multi-tenant DCs.

From the perspective of the policy-based DC network, the advantages of NUAGE are mainly originated from the separated control from the infrastructure. Specifically, it includes the following advantages:

- The network slicing can be visibly controlled, and the providers, tenants, groups, and users are provided with the role-based services.
- Any existing DC is able to be virtualized and automatized.
- It is compatible with any open-software-based products, such as Openstack, CloudStack [3], VMware Cloud.
- It provides complete virtualization and automatically finalizes the connection between the internal network and DC, and the connection between DC and the enterprise's VPN.

Nuage virtualized network services (VNS) are the complementation for the existing IP VPN and Carrier Ethernet VPN. But unlike these VPN services, Nuage VNS are specifically developed for the cloud-based IT enterprise with consumption patterns. According to the demands from enterprises, Nuage VNS provide flexible and unconstrained network services matching with the dynamic cloud environment.

Traditionally, the network services purchased from the providers are tightly connected to the private network infrastructure. Hence, the services have to be modified according to the actual demands, but the cloud environment is not optimized. Nuage VNS is a novel approach to construct the wide area network for seamlessly connecting the enterprise, regardless of size or geography, while the customized networking demands are decreased. Nuage VNS is a coverage-based approach to maximize the feasibility that any available access techniques from several providers can be effectively integrated.

In Fig. 6.2, it shows Nuage architecture mainly consisting of cloud consumption, extensibility and security, flexible network, and operational scalability. Specifically, the cloud consumption can unify the control of public and private cloud for providing the following services:

- Accounting through cloud, such as OpenStack, CloudStack, etc.
- Management through open APIs, such as OpenStack Horizon, etc.
- Customization through extensive development environments, such as Kubernetes [1], Mesos [2], etc.

The extensibility and Security enables Nuage to:

- Integrate with other applications, such as database, security devices, operating systems, etc.;
- Control the network resources through policy, pre-configuration or user interface;
- Customize.

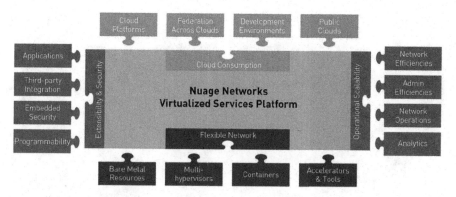

**Fig. 6.2** NUAGE architecture (*Source*: Nuage Networks)

The flexible network supports to:

- Controlling, virtualizing and managing resources, without upgrading;
- Coexistence of multiple virtualized environments.

The operational scalability provides:

- Efficient multi-tenant operation and multicast, network template, and other network functions;
- Visualized NFV.

# References

1. E.A. Brewer, Kubernetes and the path to cloud native, in *Proceedings of the Sixth ACM Symposium on Cloud Computing* (ACM, New York, 2015), p. 167
2. B. Hindman, A. Konwinski, M. Zaharia, A. Ghodsi, A.D. Joseph, R.H. Katz, S. Shenker, I. Stoica, Mesos: a platform for fine-grained resource sharing in the data center, UCBerkeley, Technical Report, UCB/EECS-2010-87 [Online]. Available: http://www.eecs.berkeley.edu/Pubs/TechRpts/2010/EECS-2010-87.html (2010)
3. R. Kumar, K. Jain, H. Maharwal, N. Jain, A. Dadhich, Apache cloudstack: open source infrastructure as a service cloud computing platform, in *Proceedings of the International Journal of Advancement in Engineering Technology, Management and Applied Science* (2014), pp. 111–116
4. L. Le, E. Hossain, Multihop cellular networks: potential gains, research challenges, and a resource allocation framework. IEEE Commun. Mag. **45**(9), 66–73 (2007)
5. X.J. Li, B.-C. Seet, P.H.J. Chong, Multihop cellular networks: technology and economics. Comput. Netw. **52**(9), 1825–1837 (2008)
6. Y.-D. Lin, Y.-C. Hsu, Multihop cellular: a new architecture for wireless communications, in *INFOCOM 2000. Nineteenth Annual Joint Conference of the IEEE Computer and Communications Societies. Proceedings. IEEE*, vol. 3 (IEEE, Tel Aviv, 2000), pp. 1273–1282
7. A. Osseiran, F. Boccardi, V. Braun, K. Kusume, P. Marsch, M. Maternia, O. Queseth, M. Schellmann, H. Schotten, H. Taoka et al., Scenarios for 5g mobile and wireless communications: the vision of the METIS project. IEEE Commun. Mag. **52**(5), 26–35 (2014)
8. H.M.P. Team, Metis: striding towards 5G. Communicate **73**, 49–52 (2014)

# Chapter 7
# 5G-Based Applications

**Abstract** This chapter introduces some representative applications based on 5G, including RAN sharing, Multi-Operator Core Network (MOCN), fixed mobile convergence, small cells, etc.

## 7.1 RAN Sharing

With deeper network sharing, especially the RAN sharing, the traditional telecommunications industry is able to transform to a deeper win-win mode. Network sharing technology has the following advantages:

- The operating expense, and 40 % capital expenditure for infrastructures are saved;
- A new profit model is provided;
- The market barriers encountered by management are reduced;
- The focus of competition transfers from network deployment cost to service innovation;
- The network deployment is accelerated;
- With green network techniques, the environmental pollution is reduced.

As a major provider of telecom industry, ZTE has developed a solution for RAN sharing, including two main components, i.e., dedicated carrier RAN sharing (logic Base Station Controller (BSC)/Radio Network Controller (RNC), and shared carrier RAN sharing.

Figure 7.1 illustrates the dedicated carrier RAN sharing, which allows the physical sharing to separate from the operator's logical control. It is independent of the terminal, which means that the operator can have own brand and subscriber, respectively, while the subscriber does not know whether the network is shared. This model allows operators to independently deploy service in the shared network. For example, operators providing run high-speed services is able to support Evolved High-Speed Packet Access ($HSPA^+$) [3], while another operator providing mobile TV services is able to deploy Multimedia Broadcast Multicast Services (MBMS) [2] on its dedicated carrier. They can set different parameters for different service for supporting different radio performances. In addition, the operator can independently

Y. Zhang, M. Chen, *Cloud Based 5G Wireless Networks*, SpringerBriefs in Computer Science, DOI 10.1007/978-3-319-47343-7_7

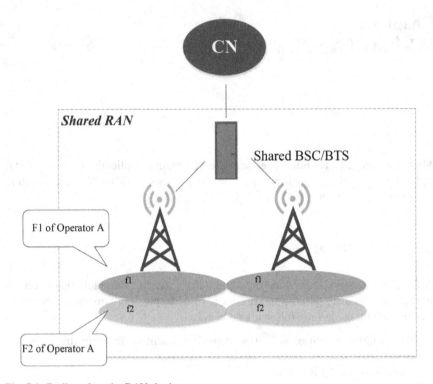

**Fig. 7.1** Dedicated carrier RAN sharing

configure the cells, such as statistical information of battery performance, fault management, and security settings, so their trade secrets are impossible to be disclosed.

Due to the simple operation, the dedicated carrier RAN sharing becomes popular in recent years. In contrast, shared carrier RAN sharing, a deeper sharing approach, can provide high spectral efficiency, but it is more complex and cannot be selected by the operator and supplier.

Figure 7.2 illustrates the solution of shared carrier RAN sharing, in which a shared element management system (EMS) is able to provide the following operation and management functions:

1. Cell-level configuration for BSC/RNC, including network hardware and transmission equipment.
2. Cite-level software upgrading and status query for BSC/RNC.
3. Cite-level paramagnetic and ferromagnetic for BSC/RNC.

In addition, this solution supports Shared Network Area (SNA). Communication networks can send SNA mapping table to RNC. After the user equipment is connected, the mapping between SNA and Temporary Mobile Subscriber Identifier (TMSI) is transmitted and stored in RNC. Thus, the mobile network is available to be seamlessly controlled in the process of the location update and access handover.

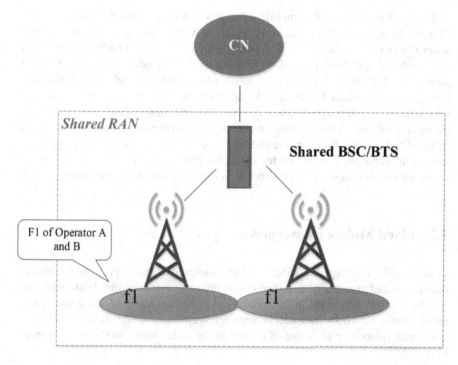

**Fig. 7.2**   Shared carrier RAN sharing

Faced with a deeper crisis in the telecommunications industry, the global operators begin to fully use the potential and advantage of RAN sharing, and actively evaluate the solutions related to RAN sharing.

## 7.2   Multi-Operator Core Network

Multi-Operator Core Network (MOCN) based on network sharing can simultane-ously connect to the core nodes from different carriers, which supports to provide the carriers with an unified wireless network [1].

Based on the demands from the operators, Huawei proposes a customized MOCN solution to satisfy the needs of sharing a wireless network between two companies, which can be deployed within 3 months. Through the network sharing supported by MOCN, the operators can share the license fees, reduce network construction costs and expand network coverage, and finally enhance the core competitiveness. The MOCN standard defined by 3GPP requires that the frequency should be shared by all the carriers. But in Huawei's customized solution, it also supports the operators have some private frequencies, i.e., the sharing and exclusive frequency can be flexibility implemented.

In a traditional network, a mobile subscriber can only access the home network of a single operator. However, in a typical MOCN scenario, mobile subscribers under the same Absolute Radio Frequency Channel Number (ARFCN) can select from multiple operators' networks. However, mobile phones are diverse and often unreliable. If a subscriber relies entirely on the handset to select networks, the wrong network might be chosen. In addition, the different processes through which network equipment handles the two aforementioned categories of mobile phones complicate address routing and O&M. By interpreting the Layer-3 messages reported by handsets, it screens the message reporting differences in both categories, forcing all handsets to select routes under the RNC's control; none is allowed to select networks on its own; this ensures registration to the correct home network [1].

## 7.3   Fixed Mobile Convergence

Fixed mobile convergence (FMC) is an emerging technology, which aims at integration and creation of a unified communication infrastructure from fixed and wireless mobile networks. In this converged communication infrastructure, users move across networks and access services seamlessly using different devices. Voice and video over IP is one of the emerging technologies in the realization of FMC [8].

- AT&T Fixed Mobile Convergence: With the growing popularity of mobile devices, it is difficult to control the wireless costs, especially without budgets. By AT&T OfficeDirect solution,[1] it is able to: (1) create the mobile routing parameters, (2) use private branch exchange (PBX) combining with AT&T wireless transmission to complete and select appropriate international telephone lines, and (3) make full use of AT&T to provide particular wireless transmission to various devices.
- NEC uMobility: With NEC's uMobility, businesses are now able to provide their employees with single number reach, unified voice messaging and enhanced in-building coverage on their mobile devices through their corporate Wi-Fi network. Employees, via their mobile device, can effortlessly roam on and off campus, from their business's Wi-Fi to cellular networks and back again.[2]

---

[1]https://www.business.att.com/enterprise/Family/mobility-services/fixed-mobile-convergence/.
[2]https://www.nec-enterprise.com/products/Business-Mobility/Fixed-Mobile-Convergence.

## 7.4  Small Cells

Small cells are low-powered radio access nodes that operate in licensed and unlicensed spectrum that have a range of 10 m to 1 or 2 km. They are "small" compared to a mobile macrocell, which may have a range of a few tens of kilometers. With mobile operators struggling to support the growth in mobile data traffic, many are using mobile data offloading as a more efficient use of radio spectrum. Small cells are a vital element to 3G data offloading, and many mobile network operators see small cells as vital to managing LTE Advanced spectrum more efficiently compared to using just macrocells.[3]

At smallcellforum,[4] it divides the use cases of small cell into the following categories:

1. **Residential small cells**: They are defined as small cells intended for home or small office applications, which are based on indoor environment, especially the location [11].
2. **Enterprise small cells**: With the great success of LTE(-A) outdoor, LTE-based small cell technology has become popular and is penetrating indoor enterprise environment, co-existing with WiFi networks, to provide better user experience or Quality-of-Experience (QoE) [4].
3. **Urban small cells**: They are defined as licensed small cells, deployed by operators in areas of high demand density on an open-access basis to all the customers of the operator; they can be deployed outdoors on street furniture; or indoor public locations such as transport hubs and retail malls [9].
4. **Rural and remote small cells**: They are expected to provide connectivity to the areas outside of towns and cities, far from existing coverage and mobile infrastructure, on-board coverage moving with users, rapidly deployable short-term coverage and limited to specific service or user group [6].

## 7.5  Other Applications

1. Mobileflow: It is a blueprint for implementing current as well as future network architectures based on a software-defined networking approach, and it enables operators to capitalize on a flow-based forwarding model and fosters a rich environment for innovation inside the mobile network [7].
2. FluidNet: It is a scalable, light-weight framework for realizing the full potential of C-RAN. FluidNet deploys a logically re-configurable front-haul to apply appropriate transmission strategies in different parts of the network and hence cater effectively to both heterogeneous user profiles and dynamic traffic load

---

[3]https://en.wikipedia.org/wiki/Small_cell#cite_note-Micro_Markets-1.
[4]http://www.smallcellforum.org/.

patterns. FluidNet's algorithms determine configurations that maximize the traffic demand satisfied on the RAN, while simultaneously optimizing the compute resource usage in the BBU pool [10].

3. Network Store: It is a revolutionary vision of 5G networks, in which SDN technologies are used for the programmability of the wireless network, and where an NFV-ready network store is provided to mobile network operators, enterprises, and over-the-top third parties. It serves as a digital distribution platform of programmable VNFs that enables 5G application use-cases, to provide a digital marketplace, gathering 5G enabling network applications and network functions, written to run on top of commodity cloud infrastructures, connected to remote radio heads [5].

# References

1. R.L. Aguiar, A. Sarma, D. Bijwaard, L. Marchetti, P. Pacyna, R. Pascotto, Pervasiveness in a competitive multi-operator environment: the daidalos project. IEEE Commun. Mag. **45**(10), 22–26 (2007)
2. M. Gruber, D. Zeller, Multimedia broadcast multicast service: new transmission schemes and related challenges. IEEE Commun. Mag. **49**(12), 176–181 (2011)
3. H. Holma, A. Toskala, K. Ranta-Aho, J. Pirskanen, High-speed packet access evolution in 3gpp release 7 [topics in radio communications]. IEEE Commun. Mag. **45**(12), 29–35 (2007)
4. U.P. Moravapalle, S. Sanadhya, A. Parate, K.-H. Kim, Pulsar: improving throughput estimation in enterprise LTE small cells, in *CoNEXT'15* (ACM, New York, 2015)
5. N. Nikaein, E. Schiller, R. Favraud, K. Katsalis, D. Stavropoulos, I. Alyafawi, Z. Zhao, T. Braun, T. Korakis, Network store: exploring slicing in future 5G networks, in *Proceedings of the 10th International Workshop on Mobility in the Evolving Internet Architecture*. MobiArch '15 (ACM, New York, 2015), pp. 8–13. ISBN: 978-1-4503-3695-6. doi:10.1145/2795381.2795390. http://doi.acm.org/10.1145/2795381.2795390
6. X. Ortiz, A. Kaul, Small cells: outdoor pico and micro markets, 3G/4G solutions for metro and rural deployments. ABI Research, vol. 5 (2011)
7. K. Pentikousis, Y. Wang, W. Hu, Mobileflow: toward software-defined mobile networks. IEEE Commun. Mag. **51**(7), 44–53 (2013)
8. M. Raj, A. Narayan, S. Datta, S.K. Das, J.K. Pathak, Fixed mobile convergence: challenges and solutions. IEEE Commun. Mag. **48**(12), 26–34 (2010)
9. R. Razavi, H. Claussen, Urban small cell deployments: impact on the network energy consumption, in *2012 IEEE Wireless Communications and Networking Conference Workshops (WCNCW)* (IEEE, Paris, 2012), pp. 47–52
10. K. Sundaresan, M.Y. Arslan, S. Singh, S. Rangarajan, S.V. Krishnamurthy, Fluidnet: a flexible cloud-based radio access network for small cells. IEEE/ACM Trans. Netw. **24**(2), 915–928 (2016)
11. J. Weitzen, M. Li, E. Anderland, V. Eyuboglu, Large-scale deployment of residential small cells. Proc. IEEE **101**(11), 2367–2380 (2013)

# Chapter 8
# Conclusion

**Abstract** Although there are already some 5G-relevant documents defining the technical specifications, the research on 5G is still at its initial stage. Looking into the future development of computer, network and communication technologies, 5G is expected to be the future architecture of wireless networks aiming to build a virtual, configurable and intelligent mobile communication systems.

At present, research on 5G is still at its initial stage, there are already some 5G-relevant documents defining the technical specifications [1, 3, 7]. In addition, although some researchers have discussed how to construct the 5G network from multiple perspectives, such as air interface [2], millimeter wave [4, 6], and energy consumption [5], many of these studies focus on technical details, rarely constructing the whole system from the global perspective. It can be predicted that 5G cannot be defined by a service or a typical technology. Looking into the future development of computer, network and communication technologies, 5G is expected to be the future architecture of wireless networks aiming to build a virtual, configurable, and intelligent mobile communication systems.

Along with the ongoing enhancements in bandwidth and capacity of wireless mobile communication system and rapid development of applications of mobile Internet for personal usage and business, mobile communications-related industries are transforming to a diverse ecosystem. 5G is not just an air interface technology providing higher data rates, greater bandwidth and capacity, but is a system to accommodate different business-oriented applications. Specifically, 5G should meet the following requirements and their associated challenges:

- *Sufficiency*: As users rely on mobile applications, the next generation wireless mobile networks should provide sufficient rate and capacity for users. It can be expected from the current business perspective that most of the mobile terminals need to reach 10 Mbps data rate to support Full-HD video compression. In some special scenarios, wireless terminals are required to achieve 10 Gbps transfer rate, e.g., for instant and highly fast downloads of files from a nearby access point.
- *Friendly*: Ubiquitous coverage and stable communication quality are basic requirements of a user-friendly communication system. Existing mobile communication systems cover almost all of the populated areas but still have blind spots.

© The Author(s) 2016
Y. Zhang, M. Chen, *Cloud Based 5G Wireless Networks*, SpringerBriefs
in Computer Science, DOI 10.1007/978-3-319-47343-7_8

Wireless communications to fast moving vehicles (e.g., high-speed trains) are not stable and reliable yet. Future mobile communication systems will combine a variety of means of communications, to provide users with ubiquitous coverage and reliable communication quality. 5G networks need to provide users with always-on experience to avoid connectionless and information transfer delay. Functionally, in addition to basic communication capabilities coupled with a more colorful video game entertainment, the 5G network is capable of providing richer business applications, bringing convenience to working and improving quality of life.

- *Usability*: Although 5G technology system may become complex, from the user's point of view, it is supposed to be simple and convenient as access technology will be transparent to users and terminals will be seamlessly switching between access technologies.
- *Economy*: 5G systems are supposed to be cost-efficient for users. Cost-efficiency can be achieved as the cost of investment into the infrastructure will be reduced and network resources will be more efficiently utilized.
- *Personalized*: 5G mobile systems should be people-oriented, and provide user-centric experience. Users can customize their services according to their individual preferences, and enjoy personalized services. According to the user's network environment, network service providers can provide optimal network access functions. Meanwhile, according to the user's physical environment and personal preferences, application service providers can offer personalized recommendation service.

In essence, to cope with the new requirements of 5G, such as higher capacity and data rate, support of high number of connected devices, higher reliability, larger versatility and support of application domain specific topologies, new concepts and design approaches are needed. Some existing techniques can be implemented for increasing bandwidth and ensuring more efficient transmission, for interference management and also for interworking with other systems. In addition, advances in terminals and receivers will be needed to optimize network performances. Cloud-based architecture is an interesting paradigm for 5G, together with cloud computing, MIMO, NFV, SDN, and big data.

# References

1. A. Gohil, H. Modi, S.K. Patel, 5G technology of mobile communication: a survey, in *2013 International Conference on Intelligent Systems and Signal Processing (ISSP)* (IEEE, Gujarat, 2013), pp. 288–292
2. S.G. Larew, T.A. Thomas, M. Cudak, A. Ghosh, Air interface design and ray tracing study for 5G millimeter wave communications, in *2013 IEEE Globecom Workshops (GC Wkshps)* (IEEE, Atlanta, 2013), pp. 117–122
3. Q.C. Li, H. Niu, A.T. Papathanassiou, G. Wu, 5G network capacity: key elements and technologies. IEEE Veh. Technol. Mag. **9**(1), 71–78 (2014)

4. G.R. MacCartney, J. Zhang, S. Nie, T.S. Rappaport, Path loss models for 5G millimeter wave propagation channels in urban microcells, in *2013 IEEE Global Communications Conference (GLOBECOM)* (IEEE, Atlanta, 2013), pp. 3948–3953

5. M. Olsson, C. Cavdar, P. Frenger, S. Tombaz, D. Sabella, R. Jäntti, 5GrEEn: towards green 5G mobile networks, in *2013 IEEE 9th International Conference on Wireless and Mobile Computing, Networking and Communications (WiMob)* (Lyon, 2013), pp. 212–216

6. T.S. Rappaport, S. Sun, R. Mayzus, H. Zhao, Y. Azar, K. Wang, G.N. Wong, J.K. Schulz, M. Samimi, F. Gutierrez, Millimeter wave mobile communications for 5G cellular: it will work! IEEE Access **1**, 335–349 (2013)

7. A. Tudzarov, T. Janevski, Functional architecture for 5G mobile networks. Int. J. Adv. Sci. Technol. **32**, 65–78 (2011)

Printed in the United States
By Bookmasters